HOPPING DOWN IN KENT

By the same author

KENT VILLAGES

HOPPING DOWN IN KENT

ALAN BIGNELL

ROBERT HALE · LONDON

First published in Great Britain 1977
Reprinted 1980

ISBN 0 7091 6257 X

Robert Hale Limited
Clerkenwell House
Clerkenwell Green
London EC1

Printed in Great Britain by
Clarke, Doble & Brendon Ltd,
Plymouth and London

CONTENTS

ILLUSTRATIONS

ACKNOWLEDGEMENTS

Illustrations 9 and 18: Radio Times Hulton Library. Remainder supplied by the author courtesy of the *Kent Messenger*. Illustrations 1, 2 and 4 by permission of Kent County Archives

ONE

Hopping Was How You Remember It

When you go a-'oppin',
'Oppin' dahn in Kent,
See ol' Muvver Riley
Livin' in a tent—
Wiv a E I O, E I O, E I E I O!

That song never was a top pop. At least it never made the charts. Well, it wouldn't. It was probably never heard outside Kent and a small part of south-east London and there only for a month or so in any one year.

Yet, forty years ago, it was sung in hop gardens all over Kent by Cockney choirs of anything from a couple of dozen to a hundred or more voices, more enthusiastic than musical, welling up spontaneously and lasting for as many verses as the singers' ingenuity could sustain before it died away to be replaced by some other ditty of similar inanity and universal popularity.

Today, it is just one of the many echoes of a past that hangs wistfully around those same hop gardens when the tractors go among them and a few dozen people cut the bines and feed them into machines.

Nostalgia for a pastoral past becomes tedious, and anyway almost always it recalls only part of a story, the whole of which may be very different. Yet, for a fact, of every hundred people who remember the hey-day of hopping down in Kent, fifty or sixty years ago, only one or two do so with any sort of thankfulness that that day has been mechanized into history.

Hopping was always much more than just the harvesting of hops. It was the season itself, and every year, at about the time when summer begins to think of migrating south in the wake of the swallows, Kentish folk who have never picked a hop in their lives still chafe the changing season's chill from their fingers and give the sharp-edged mists of late summer's early mornings a grudging welcome: "Ah—a proper hopping morning, this!"

The old style, pre-mechanization hopping was a four or five week holiday from normality during which the sights and sounds and scents of hop-picking spilled over from the hop gardens into the villages and the towns of Kent, and brought a new dimension to county life.

Of course, Kent never was the only hopping county in Britain. Despite their relative delicacy and the fact that they are prone to a variety of infections and infestations, many of which can be disastrous or even fatal to the crop, hops do adapt pretty well to many different climates and soil conditions. They are native to many different parts of the world and, at one time or another, they have been grown in almost all the English counties, with varying degrees of success. They grow best on protected sites, where the soil is strong, well-drained and loamy. In England, that has meant, in the main, Kent, Sussex, Hampshire, Hereford and Worcestershire, with, for a time, some plantations in Norfolk, Suffolk and the rest of East Anglia.

What made Kent the premier British hop-growing county, where far more than half the total national hop crop came from, was the combination of suitable sites, suitable soil and suitable economic conditions, giving rise over the years to a concentration of expertise that others could not rival.

Today, it comes as something of a shock to many Kentish men (and Men of Kent, too, of course) to have to remind themselves from time to time that hops are grown elsewhere, too. That illusion of exclusiveness is supported by the curious custom of other counties of growing their hops in hop yards whereas, as everyone in Kent knows, hops grow in gardens.

And, certainly, for the thousands of Londoners who still remember fondly their own hopping days, the only kind of hopping they ever mean is that which they came to, year after year, "dahn in Kent".

Even in Kent, though, the hops have settled themselves in two fairly distinct areas. In East Kent, they grow mainly around Chartham and Canterbury, with an important straggle of gardens along the North Kent coast around Faversham and Sittingbourne. In any of these areas, growers will assert that it is here that the very best of all Kent hops are grown.

Unfortunately for the honest seeker after truth, all through the Weald, Mid-Kent and South-West Kent, growers tell anyone who enquires that the varieties grown there are superior to all others.

The most extensive area of Kent hop gardens extends through Mid-Kent, around Maidstone and Tonbridge, south to the Kent Ditch boundary with Sussex where it spreads from Edenbridge and Tunbridge Wells eastward to Hawkhurst and Tenterden.

There is a small West Kent hop-growing region around Sevenoaks which once extended into the then rural districts of Bromley and Bexley, but these London boroughs have, this century, fattened upon people and homes until there is very little agriculture left in them, and their residents had to join the south-east London migrants if they wanted to carry on their own individual hop-picking traditions.

Throughout the county generally, hops and fruit tend to grow side by side, and all through the Low and High Weald fruit-growing region, hops represent about eight per cent of all the crops grown.

Hopping down in Kent was a social phenomenon that, for at least 200 years, was unique in all England. For most of that time it brought, in two or three days, tens of thousands of some of the poorest people in the country into the Kentish countryside. For perhaps four weeks, it concentrated them into a few hundred acres and then, hopping over, it packed them off home again and thought very little about them until the same time next year.

Perhaps George Orwell was right when he supposed the villages probably benefited from this annual injection of outside life. Certainly to the Kent local authorities it was an administrative problem that became more complex as passing time honed the social conscience into a goad with which to prod them into accepting an ever larger share of the responsibility for the welfare of these visitors.

Tantalizingly, we know very little of what the hoppers who came to Kent from London during the first part of the nineteenth century or before that thought about hopping or what it meant to them. We have no more than hints, often obscured more than enlightened by the interpretations of those who bequeathed them to us. Most of the hoppers were only semiliterate, with little ability or inclination to express themselves in writing in any form that survives today.

Until the nineteenth century, no one cared very much what the labourers thought of their labours, so although we can glean a good deal about the growing of hops before that, and even about the physical conditions in and around the hop gardens of those days, we know almost nothing of the thoughts and opinions of the people who picked the hops.

During the nineteenth century, hoppers were often referred to by others as poor, unhappy—by implication, indeed, almost sub-human—creatures, depraved rather than deprived, in whom the recognition of any virtues at all amid the pervading vices revealed more compassionate discernment in the observers than any redeeming characteristics of the hoppers.

Yet, almost as often, the hoppers were depicted as simple, happy people, delighting in the healthy rustic pursuit that was the popular poetical view of almost all rural activities anyway.

The truth, as is its invariable habit, lurks between the two, and is very hard to pin down. Undeniably, the hospitality that Kent's farmers and villagers extended to the Londoners all through the nineteenth and into the twentieth centuries was, for the most part, less than generous. Conditions on the farms always tended to lag well behind acceptable minimum standards elsewhere at the same time. Even in the 1950s, when the hopper huts were already beginning to be called chalets, they were pretty primitive really. Thirty or forty years earlier, many of them were nothing short of squalid, yet still they were an improvement on the make-shift shake-downs provided for (or, more commonly, haphazardly available to) the hoppers of fifty years before that.

The whole history of hopping down in Kent is corrugated with little outbursts of passion—strikes, fights, demonstrations, occasionally riots—always on a fairly small, localized scale, but at the same time a constant symptom of discontent and simmering revolt.

Yet that history is also decked out with reminders that these same people could change very quickly from sullen malcontents or impassioned would-be revolutionaries, into apparently carefree, happy holidaymakers, at least superficially as well satisfied with the good things that hopping had to offer as ever they were angered by its less attractive features. Time and time again, reformers found themselves thwarted in their attempts to champion the hoppers by completely misunderstanding the priorities of the hoppers themselves.

If we know so little about how the hoppers really felt about hopping, we know rather better what the local people in the Kent hop country thought about the hoppers.

Of course, Weald and East Kent farmers always relied for most of their seasonal labour upon the market towns of Maidstone, Tonbridge and Ashford. The Medway Towns supplied many of the North Kent hoppers and in the years before the dramatic increases in hop acreages during the nineteenth century, most of the Londoners had found their work in the hop gardens around Bromley and other areas soon to be embraced by the London sprawl, but at that time still firmly rural.

It was only after all the available work had been allotted in these areas, where the Londoners were to a far greater extent the 'locals', that they drifted deeper into West Kent and the Weald. Gradually, as the hop-growing boom developed, more and more of them were needed in these more distant parts of the county. On the small farms, where hops still represented a fairly small proportion of the total acreage of crops, local home-dwelling labour still met most of the needs. The Londoners made up the numbers. But on the really big farms, most of the pickers were Londoners, encouraged into Kent for the purpose. As more Londoners arrived each year, so the numbers of local people who had never been hop-picking became less likely ever to want to try it.

The truth was that, by and large, the country townsfolk felt themselves to be superior to the Londoners. The class structure of the nineteenth and twentieth centuries was the same in the city as it was in the towns, but lower densities of population in the more scattered rural communities meant that the very poor were less in evidence there. They were never *en masse* to the same extent that they were in parts of London, and when those parts of London decamped into Kent each year, comparatively few of the people in the market town centres of the hopping countryside could or wanted to identify with them.

They brought with them the generations-old stigma that dogged the poor right up to the 1930s and beyond: that of being victims of their own feckless intemperance or some other equally intolerable character defect.

So, when they arrived in Kent in their noisy, often noisome multitudes, local people took precautions against various kinds of contamination. Shop-keepers put their wares behind chicken wire; publicans put away their good glasses and brought out the cut down beer bottles that were plenty good enough for the hoppers. Cottagers were careful—as they had no need to be throughout the rest of the year—to lock their doors and close their windows when they left home, and farmers resigned themselves to an undeclared open season on practically all growing crops and a good deal of livestock as well. And everywhere, there tended to be a stronger-than-usual smell of carbolic and other parasite deterrents surrounding the local people like some kind of charm while the Londoners remained their guests.

Warranted or not, the Londoners brought down firm social barriers around the hop gardens and, except for the villagers, who regarded all outsiders, whether they came from the nearby towns or from London itself, as necessary misfortunes at hopping time and never thought seriously of not seeing their whole-year labours in the hop gardens through to their logical end, Kent people generally looked upon hopping as a colourful bit of folksiness they were glad was nothing to do with them. Most of them kept well away from the hop gardens. Those that actually went hop-picking were regarded either as unfortunate to be poor enough to have to sink to that, or as incautiously quaint.

The First World War broke down some of these barriers as a few Kent townsfolk answered the calls for help from the farmers and found that hop-picking was not as all bad as they had always supposed. The Second World War broke down more of the barriers, not just because more answered the calls and so more found they enjoyed it, but also because class barriers were breaking down anyway. Besides, hop-picking really was not as bad as it had been by then, anyway.

Even so, it was not unknown for adventurous try-anything-oncers to take a half-bin at a farm they thought would be a safe distance from friends and neighbours to whom they explained that they were "helping a friend" or "doing some social work" rather than admit they were hop-picking. Others let it be known that their doctors had advised them to go hop-picking, "just once, for the good of my health, you know!" The old hands were not fooled for a moment and anyone who was judged not to be pulling his weight at the binside was told, to the great amusement of other old hands: "Come on, nah—yer not 'ere fer the good o' yer 'ealf, y'know!"

It was the Second World War that took most of the social stigma out of hop-picking, and brought about noticeable changes in the relationship between the Londoners and the locals. By then, the Londoners were no longer the demonstrably poverty-stricken people they had been fifty years before. Many of them came in their own cars—well, their own lorries, anyway—and were well dressed. Besides, they were heroes of the blitz, every one, and deserving a certain respect on that score.

By the time the war ended, hopping was respectable and it was acceptable to admit that it could be very enjoyable. Only by then it was too late.

Of course, in the hop gardens, as distinct from the hopper camps, one hop-picking was very like another from first to last. A sixteenth-century hop-picker, reborn into the first half of the twentieth century, would have found no great difficulty about taking a place at a binside and helping with the hop harvesting.

He would no doubt have still picked as many of the small leaves surrounding the hops as he thought he could get away with, in order to increase the bulk of hops in the bin and boost his tally, without attracting too much attention from the measurer. He would still have crumbled clods of dry Wealden clay between his fingers to remove the worst of the thick black stain left by the hops after a day's picking.

The wares of the itinerant tradespeople, trudging across the wake of the bins with their baskets of buns or trays of sweets for the children might have puzzled him momentarily, but the traders themselves would have been familiar. He would have recognized readily enough the sing-song chant of the measurer as he counted out the bushes from the bins; identified the scents of the wood fires, of dry earth and wet bines and of the hops themselves. He would still have shivered in the early morning mists, sweated in the heat of the afternoon sun, stumbled over the same dusty fields, sung similar (well, not all that different!) songs, and between times allowed his consciousness to drift away into that special kind of day-dream state known to all whose hands do deft work without the need for any accompanying mental effort.

The horses and the wagons, the hop-pokes and the hop-poles, the bins and the bin-men—almost everything that was hopping down in Kent in the hop gardens themselves remained virtually unchanged for three hundred years. Even the people, for all their styles of clothing changed and made them think they did, too, really changed very little.

Perhaps hoppers always came to Kent with mixed feelings. No doubt the people of Kent always received them with mixed feelings, too. But one thing is certain: in every hundred people alive today who have ever been hopping, one or two will want to forget the experience, but the rest will want to remember. The least prompting by a newspaper or magazine article or a radio talk or a chance encounter over a pint of beer always brings the memories flooding back. They remember the journey to the hop garden, the settling into the hopper-huts, filling their mattresses with fresh straw and topping them off, as often as not, with a huge feather bed brought from home.

They remember wallpapering the hut walls, stowing away the belongings brought with them, setting up the oil stove that was going to provide heat for cooking, for boiling water and for drying clothes if necessary for the next two or three weeks. They remember how Mum always cut a handy length of stripped bine to hang on the horns of the bin and with which to discipline the children if they stepped out of line; how the children had to pick a pre-determined quota of hops into a box or an upturned umbrella before being allowed to leave the bin-side, and go and play with other children, or to earn the daily halfpenny for sweets from the lolly man when he came to the garden.

The bitter taste of the hop-stain on their fingers when they forgot to tear off a bit of greaseproof wrapping paper in which to hold their sandwiches, the way the black stain turned the water green when it was scrubbed off at the end of the day with soap and pumice-stone, the smell of the little fires of bits of dry stick and hop leaves kindled well back from the bin at midday, and the taste of the tea made from water boiled in a tin kettle on those fires—these are the things they always remember. They always recall the evenings round the bigger fires outside the huts when, comforted with stew or pudding, they joined in the singing, perhaps accompanied upon a mouth-organ or accordion, or other evenings at the village pub.

There was always a special excitement about the end of hopping feast round the bonfire, the paying-off, the trip into town to be kitted out for the winter, the train journey home again, and the sense of sadness when, at last, it really was all over for another year.

They do not forget that hopping was hard work and very tiring; that they were often stung or bitten by insects, scratched by bines, wet, cold, hot and sore.

"Ah, but they were good times," they say—and if that is how they remember the old-style hopping down in Kent then that, for them at any rate, is how it was.

There is still hopping down in Kent. The hops are still picked, even if most of the picking is now done by machines. But for most of the old-timers, hopping—what they mean by hopping—is over. When it ended, in the middle of the twentieth century, it closed a chapter of Kent history that was quite unlike any chapter of any history of any other county in all England.

TWO

The Early Years

"Away with all wine-drinkers,
And such new-fangled thinkers,
And may they still be shrinkers,
From all good men and true."
Thus said the jovial Man of Kent,
As through his golden hops he went,
With sturdy limbs and brow unbent,
When Autumn's sky was blue.

The hop that swings so lightly,
The hop that glows so brightly,
Will sure be honoured rightly,
By all good men and true.
Let Frenchmen boast their straggling vine,
Which gives them draughts of meagre wine,
It cannot match this plant of mine,
When Autumn's sky is blue.

Part of an Old Kentish song

Can you imagine a Kentish countryside without the familiar dunce-capped kilns of mottled Kent tiles, with their whimple-white wooden cowls vaned to get the best from whatever winds might blow?

Probably not. The sight is as much of Kent as the Tower of London is of its setting. Oasts belong in Kent, and the rest of Britain and half the world as well knows Kent by its oast houses. Well, no wonder. They have dotted the Kentish countryside for about a couple of hundred years, after all. Even so, compared with the hops that brought them, they are newcomers to the landscape. There was a time when the only hops in Kent were little pea-sized weeds growing wild in the hedgerows, of value only—if, indeed, at all—to herbalists.

The origins of hops are lost in the enormity of time. There is some evidence to suggest that they were used by Babylonian brewers three thousand years ago, although the first recorded mention of a hop plant dates only from the sixth century B.C.

The Roman naturalist, Pliny (A.D. 23–79) told of hops growing in gardens as a weed around willows, and certainly the Romans used the plant as a vegetable, cutting the young shoots of the bines when they emerged above ground during the summer.

They ate them much as we eat asparagus today. So it is quite probable that the first hops in Britain were those brought here by the Romans. Whether or not the Romans actually cultivated hops on a scale larger than that which would have brought the occasional plant to a villa garden we do not know for sure. Nor can we be certain if a variety of hop grew in these islands naturally before the Romans came. Certainly, after they left us, there seems to have been no deliberate cultivation of hops here until about five hundred years later.

The first record of cultivated European hops dates from the eighth century A.D., when they were being accepted as rent in parts of Central Europe, and a hop garden was mentioned in a deed of gift by Pepin, father of Charlemagne.

The esteem in which the hop was held in fifteenth-century Europe may be demonstrated by the fact that it was the insignia of an order of chivalry, *l'Ordre du Houblon*, founded by John the Fearless, Duke of Burgundy. John was cousin of the mad King Charles VI of France, and from 1404 to 1419 he ruled over large areas of France after he returned from an unsuccessful crusade in 1396, during which he was held prisoner by the Turks.

Hops may well have been used by brewers in Britain before the middle of the fifteenth century, and it is commonly asserted that Parliament was petitioned, as early as 1440, to ban its use for this purpose, although this little bit of brewing history is open to dispute. We do know, though, that an English dictionary, *Promptorium Parvulorum*, of the mid-fifteenth century included an entry which read: "Hoppe: sede for beyre", suggesting not only that hops were at least comparatively well known here at that time, but that they were probably fairly commonly used "for beyre" (beer), too.

Probably, though, they were used only as one of a variety of plants used by brewers to flavour their beer. Right up to the mid-sixteenth century, the common brew of Britain was ale, sweet, malty, thick and strong. It was often seasoned or flavoured with spices of one kind or another, and common ingredients were ale-hoof, broom, hore-hound, gentian, wormwood or ground ivy. Very possibly hops or the seeds of hops were also used where they were to be found growing wild, and they may even have been encouraged—rather than truly cultivated—by some brewers lucky enough to have a plant available, as a kind of seasonal speciality.

There is no doubt, though, that the deliberate cultivation of hops as a farm crop was introduced into England by the Flemish weavers brought to England, and particularly to Kent, by Edward III during the fourteenth century.

It was these weavers who brought with them their taste for the lighter, clearer Continental beer which had, for some time, been made with hops. Rather than import all their beer, or their hops, they taught the local farmers to grow hops so that they could brew their own.

The farmers, quick to recognize that here was a crop that grew well and brought them a good return for their capital outlay, learned the necessary skills quickly. At this time, remember, most English farming was still very much on a subsistence level. Farmers grew what crops and raised what stock they needed for their own families and could sell or exchange for labour to neighbouring villagers. Kent, more than most counties, early learned the value of cash crops that could be sold in the London markets as well, but generally speaking it was their own surplus that went to market, rather than large quantities of specially grown market crops.

Hops quickly demonstrated their worth as a cash crop, much of which was grown only to be sold. It was, therefore, a very profitable crop, in spite of the very big element of good luck and skilful husbandry necessary to get a crop at all.

So profitable, in fact, that soon laws were needed to prevent farmers from upsetting the whole national economy by abandoning arable farming in favour of hops. An old rhyme dating from 1424 tells us: "Turkeys, Carp, Hops, Pickerel [pike] and Beer, Came to England all in a year." There are several variations on this theme naming different things, but they all include hops, which may be regarded as significant.

Hop plants began to appear in illustrations for books in the fifteenth century, too. A woodcut printed by Peter Schoyffer at Mayence in 1484 showed a hop plant, and another rather more easily recognizable woodcut illustration appeared in a book published in Venice in 1499.

Certainly by the sixteenth century Henry VIII was under some pressure from the herbalists of his realm to forbid the use of hops which they saw as a threat to their trade. At that time ale was commonly flavoured with cloves and cinnamon which, of course, was supplied to the brewers by the herbalists. But Henry's objections to hopped ale may have reflected his personal taste, too. In the records of his household at Eltham in 1530 there is an instruction to his brewer not to put hops or brimstone into the ale. Whether this was because Henry did, in fact, prefer his ale to have traditional English qualities, or whether it was because he was persuaded by critics of the hop that it was actually harmful to the drink, we do not know.

There can be no certainty, now, about where the first hops were grown, as a deliberate crop, in England. Kent claims that distinction; more specifically, tradition puts the first hop gardens in the parish of Westbere, near Canterbury, in 1520. Writing in 1717, Daniel Defoe described the districts around Maidstone and Canterbury as "the Mother of Hop Grounds in England", but an area as vague as that could take in practically all the Kent hop-growing areas anyway, so it is not a great help in tracking down the elusive first Kent hop garden.

Another claimant for the same title is Little Chart, which is near Ashford. There the garden for which the title was claimed ended its career in 1976. Parish records suggested that an area there had been under hops since, in the reign of Henry VIII, grapes were dug up to make room for them. Traditionally, the 400-year-old hop garden was on the right-hand side of the Charing–Pluckley road, opposite Chart Court. But in 1976, the owner of Chart Court and the old hop garden announced his intention to turn it over to strawberries, partly in protest against the effects on British hop-growing of the Common Market insistence on seedless hops.

Wherever and whenever hops were first grown in this country, certainly J. M. Russell in his History of Maidstone says that in 1426 information was laid against a person at Maidstone for "putting into beer an unwholesome weed called a hoppe". This is probably better evidence that wild hops were used as one of many alternatives for flavouring ale before cultivated hops were grown here specifically for clearing and preserving beer.

Hops began to be an important crop in the sixteenth century. It was easier to establish them in Kent, where the enclosed field system of farming was already old than in the traditional open-field systems elsewhere. There was plenty of wood for poles in Kent, too, and London provided a good supply of the necessary seasonal labour. Hoppers began to come to Kent from London at least from the mid-seventeenth century. London also provided a good market for the hops once they were picked, because brewing was already going on there on an industrial scale.

Another feature that favoured Kent for England's major hop-growing county was that Kent farmers were already comparatively more prosperous than many of their neighbours in other counties. They were therefore more able to speculate the high cost of initial outlay on poles, labour, and so on.

Poles had to be replaced quite frequently, and were a major item in the costs of growing hops. It was only as recently as about a hundred years ago that they began to be treated with creosote to preserve them against rotting.

Documentary evidence about hops is sparse among local records. Estate maps show there were small areas of hops on many farms, but hops, like other crops, were generally regarded as part of the land itself for inheritance purposes, and so tended not to be listed in inventories of possessions. In many cases, hints about the extent of a holding's proportion of hops have to be gleaned from lists that include such capital items as hop poles, oast houses, and so on.

However, in 1524, a decree was published forbidding the use of hops to embitter beer on the grounds that they "tend to make the people melancholy", and it was because of the outcry that was beginning to be raised against "this wicked weed" that some old writers dated the arrival of hop cultivation in England from this date.

Once it was here, though, the hop's popularity went up and down for many years before it became generally accepted. Henry VIII forbade it in brewing. In Queen Mary's day, Catholics condemned hops simply because they came from the Protestant Low Countries.

Elizabeth I encouraged the farming of hops, but by the early

1600s, Parliament was petitioning against "this wicked weed that will spoil the drink and endanger lives".

In fact, for something like a hundred years, hops were about as controversial as the so-called 'soft' drugs such as cannabis are today. There were those who regarded hops as pleasantly harmless; others who thought of them as weeds and a drug. There were even those who were prepared to blame the use of hops in the drink for inciting followers of Jack Cade's rebellion in 1450.

The first hops were almost certainly grown here by, or at any rate under the direction of, the Netherland farmers who came with or followed close behind the Flemish weavers. They, of course, already knew all about hop cultivation, whereas English farmers did not, and had to learn from or be taught by those who did. There is a record of the payment of money to one Peter de Wolf and some of his workmen in April 1550 for "planting and setting of hoppes".

But the uncertainty of their reception by English authority must have made many farmers cautious at first, and there were no doubt many experts prepared to say that hop-growing would never catch on.

That may have been why Thomas Tusser, in his book *One Hundreth Points of Good Husbandry* published in 1557, made no mention at all of hops. Sixteen years later, though, in the sequel, *Five Hundreth Points of Good Husbandry* published in 1573, he had apparently decided that hops were at least sufficiently well established to deserve a mention. In a reference under the heading:

> "A Lesson of Hop-yards," Tusser wrote: "Where hops will grow, Here learn to know, Hops many will come, In a rood of Room. Hops hate the land With gravel and sand. The rotten mould For hop is worth gold. The sun south west For hop-yard is best. Hop-plot once found, Now dig the ground. Hops favoureth malt, Hops thrift doth exalt. Of hops more reed, When time shall need."

In 1574 Reynolde (or Reginald) Scot, Sheriff of Kent, published his *A Perfite Platforme of a Hoppe Garden.* Hops had well and truly arrived in England, and especially in Kent, by then. Scot was a Man of Kent: one of the Smeeth (near Ashford) Scots, who learned what he knew about hops from working with and among them. In the course of a quite detailed exposition about hop husbandry generally, he had this to say "On the gathering of Hoppes":

> "... what time your hops begin to change colour [that is to say] somewhat before Michaelmas [October 11] ... you must gather them and for the speedier dispatch thereof procure as much help as you can, taking the advantage of fair weather and note that you were better to gather them to rathe [too early] than to late."

He went on to instruct the grower in the method of harvesting. Pull down four hills standing together in the middle of the garden, he advised. Cut the roots off all those hills, cut the plot and level it, throw water on it, tread it, sweep it to make a "faire floore whereon the Hoppes must lie to be picked".

Then, he said, cut the stalks close by the top of the hills and remove the bine from the pole with a forked stick, pushing the bine upwards and off the top of the pole. He went so far as to point out: "For the better dooing hereof, it is very necessarye that your Poales be streight without scrags or knobbes".

He advised against cutting more bines than could be carried away within an hour or two because, if the sun was too hot, or if it should happen to rain, the cut hops would be "much impaired".

All of this, of course, was advice that would have been given by Kentish hop growers four hundred years later, almost without alteration. Scot's picture of hop-picking is a simple one. "Let all such as helpe you stand rounde about the floore and ... speedily strip them [the bines] into Baskets prepared ready before."

This reference to baskets for receiving the picked hops may indicate that baskets were used before bins were developed, or it may only mean that, even then, East Kent (where Scot gained his experience) favoured baskets while other parts of the county developed the characteristic bins.

In 1577, William Harrison wrote: "Hops in time past were plentiful in this land; afterwards also their maintenance did cease; and now being revived, where can better be found." Whether he meant hops were cultivated here anciently (perhaps by the Romans or their Saxon and Jutish successors) and then neglected, or whether he was referring to an abundance of wild hops which had come to be regarded as a nuisance and had been reduced to comparative rarity, we do not know. Perhaps his "time past" referred to no more than a century or so, and hinted at a brief spell of popularity for the hop cultivated as a kind of garden spice which, because of the opposition to the use of hops in brewing was discontinued until it was revived during the sixteenth-century hop boom.

Whichever, there seems to be no doubt at all that once hop-growing came to Kent it found a part of the world particularly well suited to it, where it could flourish and delight the hearts of growers and drinkers alike.

Harrison himself wrote in his *Description of England* (1577):

> "Of late years we have found and taken up a great trade in planting of hops, whereof our more hitherto and unprofitable grounds do yield such plenty and increase that there are few farmers or occupiers in the country which have not gardens and hops growing of their own, and those far better than do come from Flanders unto us."

It was about this time the country finally made up its mind that hops were not only acceptable but decidedly to be welcomed.

We know very little about the harvesting of those early hops in Kent. Certainly, there was none of the air of festival that gathered around the season in the late nineteenth and twentieth centuries. A woodcut picture from Scot's *The Perfite Platforme of a Hoppe Garden* shows pole-pullers at work at a hill of hops, with three poles to the hill. The practice at that time was to cut the bine and strip it from the standing pole. To do this, the pole pullers used a long pole with a curved blade that had a spike on the heel of it. It was a tool that had changed very little right up to the time the machines took over in the 1950s and 1960s.

There seems to have been a difference of opinion in those days about whether it was better to pick the hops thus stripped from the poles straight away, in the hop garden, or to take them on the bines to the 'drying house' (later to be known as the oast) to be picked there. The advantage of picking them where they grew was that the hops were still crisp and came off the bines more easily. Besides that, they did not shed their seeds on the journey to the drying house. They were picked into blankets, the first forerunners of the bins, or into baskets.

The price of hops fluctuated wildly over the years. In 1583

they were making £1 6*s*. a hundredweight. By 1591, that figure was up to £2 18*s*. a hundredweight, and by the first years of the eighteenth century it had crept up by degrees to as much as £7 a hundredweight, only to drop in 1701 to £1 15*s*. before lifting again, in 1702 to £5.

But the prices varied, not only from year to year, but from region to region, and even sub-region to sub-region. For example, in 1793, the East Kent price for hops was from £11 11*s*. to £13 13*s*. a hundredweight, while in West Kent they were making £10 10*s*. to £11 10*s*. a hundredweight.

New hop gardens continued to be planted. Robert Goodsall in his book *Stede Hill* in which he traced the history of the house he made his home at Harrietsham, recorded that the mid-seventeenth-century Dr Stede "planted an extensive hop-garden", and in 1644 a tripartite indenture that included Isaac Hunt of Bethersden, yeoman, referred to ". . . one acre of hopp ground in the occupation of John Bull".

There is an interesting entry in the same book that gives some indication of values in the mid-seventeenth century. Documents probably drawn up for the Commissioners charged with sequestering estates during the Commonwealth detail land belonging to the Stedes and refer to: "Southfield land arable and meadow and ye old hop ground adjoyning, value £1 acre: 4 acres, 0 rods, 25 perches; yearly rent £4 3*s*. 2*d*."

Later in the same list is the entry: "Old hopground £1, 4 acres, 1 rod, 34 perches £4 9*s*. 0*d*. yr rent."

Throughout the fifteenth and sixteenth centuries, considerable quantities of foreign hops were imported by Southampton and other merchants. In 1608, for instance, they were buying them in for eight pounds a hundredweight. But the foreigners were becoming unacceptably fraudulent in their dealings and James I passed an Act for "avoyding of deceitfull sellinge, buyinge or spendinge currupte and unwholesome Hoppes". The Act tells of foreign exporters selling us hops in which were mixed leaves, stalks, powder, sand, straw and wood, "dross and other soile to increase weights".

Under the Act, offenders were to forfeit their hops, half the value of which was to go to the king, and half to whoever detected the fraud.

The value of hops as a crop for the farmers was undeniable. It was a very chancy crop, though, demanding a good deal of risk on the part of the grower. A healthy-looking crop could be blighted almost literally overnight, so that the whole year's husbandry went to waste. However, there is an element of gambler in every farmer everywhere and throughout all time. In seventeenth-century England, gambling instincts were especially strong, and for those who gambled on hops and won, it paid off handsomely. As Samuel Hartlib wrote in 1659: "Now it is known that Hoppes of England are the best in the world."

A pretty fair indication of the sort of prize that a hop garden offered its grower can be gleaned from what John Worlidge of Hampshire wrote in 1669:

> "An acre or two of ground so managed by one or two persons shall redound, one year with another, to more advantage than 50 acres of arable land, where there is much more time, cost and expense bestowed on it."

And the herbalist Nicholas Culpeper, in his *English Physitian* (1695) wrote: "They [hops] . . . are found in all parts of this land."

At first, much of the hop crop was used by the growers themselves, and they were often also their own brewers, too. Many farmers built drying houses—before they borrowed the Flemish name and Anglicized it into oast house—as additions to their farms and grew, picked, dried and used their own hops for their own beer. It was not long before the value of the Kentish hops as a cash crop, to be grown in quantities far greater than those needed for self-consumption and sold in the market to brewers all over England was recognized and began to be exploited.

London had a hop market from 1681, in Little Eastcheap, north of the River Thames. Later, though, this moved to The Borough, Southwark, which has remained the London hop market area ever since.

The "acre or two" referred to by Worlidge, and the four acres referred to by Goodsall soon became very small hop gardens. Before the seventeenth century was out, Celia Fiennes, traveller and diarist, was recording a journey from Sittingbourne to Canterbury in which she wrote of passing "great Hop-yards" on both sides of the road. That was in 1697, and she also recorded that "this year was great quantetyes of that fruit here in Kent".

After the ups and downs of popularity enjoyed by the hop in its early days in this country, it settled down to begin to become a typical southern—and particularly a Kentish—crop.

Not that it lost its detractors at all suddenly. Diarist John Evelyn was one of those and he deplored the waste of good ship-building timber that was used for hop-poles. He raised a petition against hop-growing on those grounds, but he was ill-informed on that particular subject. The best of the ship-building timber—oak—was not particularly favoured for hop-poles, for which farmers tended to prefer ash, chestnut or larch, none of which figured at all prominently in the ship-building industry of Evelyn's day.

Anyway, Evelyn's petition did not stop the spread of the hop, although the cost of the poles was always a major concern for growers once they had settled down to increasing their acreages of hops. It was one of the considerations that favoured stringing when that was introduced from America, via Germany, although it did not catch on as fast as it might have done because poles, although a much heavier initial cost, lasted several seasons, whereas the string had to be renewed every year.

THREE

The Eighteenth Century

Ye Farmers of Kent, who are jolly and gay,
Come listen a while, and pray mind what I say,
May this season be crowned with plentiful crops,
And off from an acre a load of good hops.

Oh! may they prove fine, too, and fetch a great price,
That ye, my brave boys, may get rich in a trice;
For, as ye are ever both hearty and free,
Success to ye all, for to fill ye with glee!

To crown your repast in the hopping this year,
I wish that the weather may be fine and clear;
For when it is wet, it is wretched and sad,
From morning till night in a hop-ground to pad.

Then to see the poor hoppers, alas! what a sight,
'Tis enough to put modesty into a fright;
For they are so draggl'd and wet to the skin
They're much to be pitied, their clothing's so thin.

In case of this weather, let there be no flaw,
Take care to provide them with plenty of straw,
That when the poor wretches retire to their nest,
They may lie in comfort and all take their rest.

But above all that's said, pray don't cheat the King;
For if you do that, it is sure a sad thing:
As he'll have his duty by hook or by crook!
Beware, oh! Beware, lest you're in the black book.

Tho' ye have more honour, at least so I trust,
(I'd have ye be always quite upright and just),
For honour and honesty carries the sway,
Then from these great maxims ne'er venture to stray.

So here ends my theme, for I've no more to say,
Only wish that each guest will attend to my lay,
For, faith, no exceptions are meant in my song,
And them that cries yes . . . by my soul, they are wrong.

Published first by the Kentish Gazette *of September, 1776, under the heading: "Hops: A New Song for the year 1776 addressed to the Farmers of Kent"*

The great hopping down in Kent tradition really only dates from the mid-nineteenth century, but long before that the Kentish hop gardens were drawing crowds of casual workers from the towns and from London for the hop-picking season each year.

Many of them, it is true, did not come for the hop-picking alone. They drifted out of the towns and into the countryside months before, in search of work on the farms: fruit picking, harvesting, hop-picking—anything that made farmers glad enough to take on all the labour they could get for the short harvest seasons of first one crop, then another. Some of them came, too, not primarily for the work they could get, but for the generally easier living that offered in the summer countryside, where potatoes and other root crops could be gouged out of unguarded land, where there was livestock to be stolen, streams to be fished, rabbits and hares to be poached.

To be sure, the penalties for being caught at any of these activities were severe. A man could be transported or even hung, according to shifting fashions of public indignation, and theft was sometimes punished far more harshly than murder. Nevertheless, the countryside in the eighteenth century was wide and comparatively empty, and in any case full of places to hide. If a man could not practice a little self-help in the countryside, he had almost no chance of survival in the towns. Most of the itinerants who journeyed to Kent in those days had served lifelong apprenticeships in self-help in the towns already.

They were never very welcome, these professional vagabonds. But they were needed, and so they were tolerated, at least by the farmers, if less readily by the villagers.

Hop-picking offered a particular opportunity for these labour-market scavengers. Hops must be picked at just the right time, and then quickly. Three or four weeks (perhaps five) was about as long as any hopping ever lasted. Then, too, hops are very small, and they grow in great numbers. Farmers who could get their fruit picked and their corn harvested by their own local or village-based labour, perhaps with the help of a few gipsies or tramps, needed a great many more hands to snatch the hops off the bines if they were not to watch the crop deteriorating and the prices in the markets falling away.

All through the seventeenth and eighteenth centuries, the quantity of hops being grown in England increased. By 1800 there were some 35,000 acres of hops, and that rose to 72,000 by 1878. Population was increasing, and so was beer consumption, especially in the rapidly expanding industrial areas, including London. So, after the local families had been recruited into the hop gardens, there was always work for a considerable corps of 'foreigners', as Kentish country folk always have called the new arrivals.

An Act of Parliament of 1710 told of "great numbers of poor and indigent persons" commonly employed in the gathering and picking of hops. At that time, they were not (as they were, for example, 150 years later) mainly from the London area. Some were, but many were gipsies and other travellers and the poor and unemployed from the towns of Kent and the rest of the south-east, as well.

In the eighteenth century, there were excise and custom duties on hops. In 1711 the excise duty was a penny a pound, and the customs duty threepence a pound. By 1778, the excise was a penny plus 5 per cent, and the customs duty £5 18*s*. 10*d*. a hundredweight.

Although it is the Weald of Kent, Mid-Kent (around Maidstone) and the south-west that the world most commonly thinks of as the great hop-growing area of the county, there has always been an East Kent hop-growing area, based on Canterbury, too. In 1724 Daniel Defoe wrote of "the most surprising increase in hop grounds" there, saying there were 6,000 acres of hops being grown within a few miles of Canterbury. Edward Hasted, the great Kent topographer, towards the end of the century, commented that around Maidstone, Faversham and Canterbury (which were, of course, virtually all the Kent hop-growing areas) the hop grounds had "increased greatly of late years".

When Pehr Kalm, the Swedish-born Finn, Professor of Economics at the University of Finland, visited Kent in 1748 as part of his tour of Europe in search of scientific information, he noted in his diary that most of the hop gardens were at Canterbury, with Maidstone as the next most important growing area. He also noted that the fertilizer most commonly used in the hop gardens then was burnt horse manure, mixed with road dust and he added that the hops needed a very rich earth. He also wrote of the hops growing from hills on tripods of poles, which must have been a variation of the four-pole method described by other writers of the same period.

But the East Kent hop gardens never attracted the same numbers of foreigners that the Weald and West Kent did. That is easily understandable. Most of the foreigners came into the county from the North and West, and, as time passed, specifically from London. They had no means of transport other than their own two feet, and so they spread across the county, like a stain spreading across a table-cloth. The farms nearest London's boundaries got their fill of the available casual labour first, and the army of itinerants moved south and east, thinning out as it went on. East Kent always had to rely more on local people and people from the villages not necessarily truly local but still in the general area of East Kent. There were always gipsies and other habitual travellers, with no permanent homes in any cities or towns, who made an annual practice of working their way to this end of the county specially for the hop-picking.

Early in the eighteenth century, hop-growing was big business indeed. Professor R. Bradley, Professor of Botany at Cambridge University, explained his book, *Riches of a Hop-Garden Explained* (published in 1725) with the words: "Wherein such Rules are laid down for the Management of the Hop, as may improve the most barren ground, from one Shilling to thirty or forty Pounds an Acre per Annum."

Good business indeed! He told his readers that "commonly about St Margaret's Day, the hops blossom, and about Lammas they bell, and when they change to a brownish colour and the parts of the bell are loose, and yield a grateful smell, the seed likewise beginning to turn brown, it is then we must make all imaginable haste to gather them, procuring at the time of pulling as many hands as you can. . . ."

Bradley wrote of "the old ways" of preparing a plot in the garden and treading a floor to pick the hops on to. But he judged the better way was to prepare "a frame 10ft. or 12ft.

Stripping bines from poles in 1574, from an illustration in Reynolde Scot's book of that date, *A Perfite Platforme of a Hoppe Garden*

Hop-picking was still almost a 'cottage industry' in some parts of Kent in the eighteenth century. In 1779 the bines were brought from the garden behind the cottage to the pickers in the cottage garden

Nineteenth-century pickers worked on bines still on the poles, laid across a stripped pole resting on the crossed ends of the bins

Hoppers at work: 1874

long and four feet wide with some coarse sacking cloth hung from it, so that pickers could stand three on each side and be supplied by one man who gathers them". In other words, a bin.

He warned: "It is not proper by any means to pull them while they are wet, either by Rain or Dews." Later, of course, that warning lost some of its significance as new methods of drying the hops were introduced.

Before the middle of the eighteenth century, many of the traditions and customs of hop-picking were already well established and remained to characterize this almost uniquely south-eastern counties harvest for the next two hundred years.

In what must be the most celebrated and is certainly the most rambling poem about hops and hop cultivation, Christopher Smart (who was born at Shipbourne in Kent in 1722 and was educated at Maidstone Grammar School) told an epic-style blank verse tale more than seven hundred lines long, part of which deals with Kentish hop culture, based on personal experience on his father's farm. His poem was called simply *The Hop Garden* and was first published in 1752. It includes passages about the hop-picking:

> "See! from the great metropolis they rush,
> The industrious vulgar! They, like prudent bees
> In Kent's wide garden road, expert to crop
> The flow'ry hop, and provident to work
> Ere winter numb their sunburnt hands and winds
> Engaol them, murmuring their gloomy cells . . .
> From these, such as appear the rest t' excell
> In strength and young agility, select.
> These shall support with vigour and address
> The bin-man's weighty office, now extract
> From the sequacious earth the pole, and now
> Unmarry from the closely clinging vine. . . ."

Another passage in the same poem tells how:

> "The cumulating mob,
> Strive for the mastery—who first may fill
> The bellying bin, and cleanest cull the hops;
> Nor ought retards, unless invited out
> By Sol's declining, and the evening's calm,
> Leander leads Letitia to the scene
> Of shade and fragrance—the exulting band
> Of pickers, male and female, seize the fair

Reluctant, and with boist'rous force and brute
By cries unmoved, they bury her i' the bin;
Nor does the youth escape—him too they seize
And in such posture place as best may serve
To hide his charmer's blushes; then with shouts
They rend the echoing air, and then from both—
So custom has ordained—a largess claim."

Two hundred years dealt lightly enough with this particular hopping tradition of dumping visitors into bins full of hops. There are plenty of people today who can tell of the time they were first up-ended into a bin, either as a kind of initiation to hopping, or as part of an end of hopping frivolity, or just because some of the hoppers decided it would be fun. The farmers knew it went on, of course, but they had to disapprove. Hops are delicate things and easily damaged.

Another long-lasting tradition of the hop-picking season was the hop supper to celebrate the end of the hopping. In the last days of old-style hopping down in Kent, in the 1950s, the hop supper was a great occasion, when all the hoppers gathered together, usually at the oast house, inside or out according to the size of the farm, number of pickers or temper of the weather, to drink barrelled beer provided by the farmer and to eat baked potatoes and sandwiches. Generally, it was a grand get-together, with sing-songs and reminiscences, before the hoppers packed up and went home until next year.

At some farms there were bands and dancing, and talent shows and competitions. Yard-of-ale-drinking contests were always popular; so were pie-eating contests. And the whole thing was usually accompanied by a huge bonfire using up any left-over cooking faggots and clearing away any rubbish the hoppers wanted to get rid of before they left.

Well, that was the 1950s; but those days were in direct descent from the hop suppers that were equally customary in the mid-eighteenth century.

Robert Goodsall, in his book *Stede Hill*, quotes one Robert Bottle, who had this to say about a hop supper in 1737:

"Mr. Turner's hop supper was at this time kept at Richard Daveses and he did give the Hoppers 1*s*. each to spend, and I was invited and by the order of Mr. Burham I was alotted to pay 6*d*. towards the first reconing, although I had my Violen and play'd to the Hoppers untill allmost 12 o'clock, which made me make this memorandum for it is something uncommon and ought never to be forgotten."

(The Richard Daves referred to may have been the licensee of one of the local pubs, and the uncommonness of the occasion almost certainly did not refer to the hop supper itself, but to Robert Bottle's part in it.)

Other hop-garden customs and practices of very early origin include that of demanding shoe-money or foot-money from visitors to the hop garden for the privilege of having their shoes wiped with a branch or bundle of hops. The money thus collected by the hoppers in any one garden usually went towards the cost of the hoppers' feast at the end of hopping and anyone who demurred about making the payment was, by tradition, up-ended into a bin full of hops.

Another tradition was the making and eating of 'huffkin', a kind of cake specially made for the hoppers' feast.

How eighteenth-century hoppers were accommodated on the farms is told in an account by William Ellis, a Hertfordshire man, writing about hop-picking in Kent in about 1740. He recorded that one Esquire Whitworth, of Town Malding (now Malling, near Maidstone) had over a hundred acres of hop ground. Mr Whitworth provided a little hut at every one or two bins, furnished with wheat straw for the pickers to lie on, and a cask of small beer "that they may not lose time in quest of drink". It was his practice to give to every picker every morning a quartern (quarter pint) of gin (as a preservative against Kentish ague). Mr Whitworth apparently paid one penny a bushel for gathering the hops, and he gave a feast at the end of the hopping, with a fat steer and all the strong beer the hoppers could drink.

This kind of open-handedness was not unusual and not entirely disinterested generosity on the part of the farmers. The hoppers came to Kent's hop gardens from far and wide—although, even then, very many of them were coming from London. When the picking was over, they departed again, and went home or into winter quarters somewhere. They felt no sort of obligation to Kent or to the farmers who employed them in Kent, and they certainly saw no reason, at the end of one season, to worry about whether or not they would return next year. They could, after all, seek work from any one of a hundred or more farmers in a comparatively small area, or even wander into another county altogether. So, if they were not treated well, they had no compunction at all about wholesale theft and vandalism before they moved on.

Besides, if one farmer did not spoil them in some way, he might find himself in difficulty about getting all the casual pickers he needed next year. So farmers found it worth their while to nurse a reputation among the pickers for generosity in some way or other, and this usually took the form of handouts of gin and beer and the end-of-season feasts to try to avoid inciting any kind of truculence among the pickers. A mob of out-of-county hoppers could be very nasty and it was well worth it to the farmers to keep their crews as happy as possible.

The eighteenth century saw a peak of gin-drinking in London and elsewhere that reached orgy proportions and killed great numbers of people every year. No-one drank water because it was neither wholesome (which was not always understood) nor particularly palatable (which was). Poor people drank beer, and brewing began to be a major industry at this time, with many very large breweries in London alone, as well as many others in other parts of the country.

Poverty was common and extreme, and nowhere more so than in the cities. Silk workers, framework knitters and stocking weavers in London were among the very poorest, many of them earning nine shillings a week at a time when it cost at least a pound a week to keep a man, his wife and three children. Women and children had to work to sustain life itself, and it was commonplace for poor families to live in one room and all to sleep in one bed.

J. Bannister, in his *Synopsis of Husbandry* in 1799 told of hop-pickers "generally made up of women of the lowest class" receiving a penny or a penny-halfpenny a bushel, with a dram of Geneva (gin) thrown in every morning, plus a feast at the end of the hop-picking.

Many farmers, though, also provided milk and other necessities, so that for those who could get to the Kentish hop gardens, the prospects were certainly no worse and often considerably better than they were for their every-day lives in London.

Local men who could turn their experience of years of hop seasons to earn themselves the reputation of being highly skilled hop dryers in the oast houses did very much better than their less specialized colleagues during the season. Their normal farm labourer wages shot up—they could double for the hopping—up to fifteen shillings a week or, perhaps, ten shillings, with board, plus an allowance of strong beer and spirits for the period. But hop drying was no sinecure of a job. Once artificial heat began to be used in the specially-built oast houses, the drying hops had to be tended constantly and carefully. Judgment about just when the hops needed to be turned or shovelled out on to the storage floor ready for pressing was acquired over years, and it demanded that a dryer should forgo all but the lightest of cat-naps, sometimes through a whole week at a time, to make sure the fires were kept at even temperature and the hops dried evenly and to just the right texture. What that degree of dryness and that proper texture was could only be determined by the eye and the nose and the touch of fingers as the hops were rubbed reflectively between them.

Only on Sunday, the day of rest when the fires were allowed to die and no hops were spread on the kiln drying floors could a dryer go home to his cottage and catch up on a little sleep. Even then, his employer was very likely to expect to see him in church at least once during the day, and Sunday night was a short one, for the day of rest ended at midnight, and the fires had to be ready for the new week's first measuring of hops on Monday.

Throughout the eighteenth century, hops increased in importance in Kent generally. Until then, much of the Weald and particularly the High Weald countryside where the Kent and Sussex border shared the same remains of ancient forest land, weaving and ironfounding had been the staple industries for centuries. This, in fact, was one of the most important industrial regions of all England. But then, conservationists began to realize that the apparently unlimited supply of timber for the foundry furnaces was not unlimited at all, and that every tree burned in the fires was one less for the ships the Navy needed for the wars of the period. First the water and then the coal of the North and Midlands became more and more important as the major source of Britain's industrial power and in a comparatively short time the industry itself moved out of the Weald and into what was soon to become known as 'The Black Country'.

Shorn of much of the industry that had made the area rich when others were still at subsistence farming level, the Weald became more and more dependent upon agriculture, with fruit and hops gradually emerging, throughout the eighteenth century, as the predominant crops. At the same time, the older fruit-growing areas of North Kent, around Faversham and Sittingbourne, also increased their acreages of hops.

By the end of the eighteenth century, the population of Britain was about eight million, and they were consuming about 300,000 hundredweights of hops a year. Most of these were still being bought by thousands of small brewers all over the country, and some were still going to home brewers. But more and more were being bought by the big breweries of London and other cities—including the new industrial cities that had taken over from industrial Kent with the development of new technologies.

In fact, eighteenth-century brewers used about twice as many hops in their recipes as we do today. No wonder so many farmers in Kent, where hops grew so splendidly and so profitably, plunged deep into the cultivation of hops.

It was late in the eighteenth century, too, that the most characteristic of all Kent countryside architectural features began to grace the rural scene. Until then, oast houses had tended to be generally barn-like structures wherever they were not simply additions to farm houses or farm-yard buildings. But towards the end of the century, the familiar conical-roofed kilns began to make their first appearances, with their white cowls pointing the direction of the wind and, incidentally, making sure their spicy cargoes spread the distinctive aroma as far as possible upon whatever breezes blew.

Bins were already in common use by then, certainly in south and west Kent, for hops to be picked into, and it was already the practice to measure the hops out of the bins into measuring baskets, and to pay the pickers so much a basket or tally.

Richard Tylden of Milstead near Sittingbourne kept detailed accounts of his farming activities in the eighteenth century, and his diary for August 29, 1743 records:

> "Began hoping and it lasted eight days. It was fine weather to pick 'em in. My hops were very fine this year and a pretty good crop. Hops sell now from 40*s*. to 54*s*. a cwt."

Tylden had all his hops picked into five bins. Four of those had a pair of pickers at one end and one picker at the other end. The fifth had four people, including the farmer's wife, at one end, and two at the other end. In addition, he noted, "Goody Middleton and her children pick'd in a basket".

There were two binmen to help the pickers, one man who "sometimes fetch'd the hops for them", and one dryer. The pickers were paid between four and eleven shillings for their eight days' labour. The binmen received twelve, at the rate of one and sixpence a day, and the dryer was paid sixteen shillings for his eight days' work. Half a day's pole-pulling earned ninepence for one man.

Altogether, Farmer Tylden's hop harvesting cost him £7 6*s*. 9½*d*., and he gathered 18 hundredweights 27 pounds of hops. He sold them for between sixpence and eightpence a pound according to quality, keeping about forty pounds for himself, and he paid seven pounds thirteen and threepence in hop duty on the crop. All of which added up to about forty pounds clear profit that year.

Although hop-growing was a very speculative business, profits could be so great one year as to offset losses in another. Sometimes, they had to.

FOUR

Nineteenth Century—First Half

Come, neighbours, away to the hop-grounds, away!
Behold the bright season invite,
Where pleasure attends on the toils of the day
And Labour is crown'd with delight.
Haste, haste, then and strip, as it bends from the pole
The fruit that gives vigour and strength to the soul:
Our hearts and our spirits to cheer
It warms and enlivens the true British Beer.

Let innocent mirth to loud harmony raise
And rapture pour forth all our songs in its praise—
'Tis the liquor we love—'tis the juice we revere
'Tis the spring of our courage—the true British Beer.

Charles Dibdin the Elder, 1777

All through the nineteenth century, many thousands of Londoners flocked out of the city each year in search of work on the Kent farms and, specifically, in the Kent hop gardens. There was no control over how many of them came. Kent had 40,000 acres of hops by the middle of the century, and every acre needed about two hundred pickers for the two or three weeks the season lasted.

The pilgrims had no way of knowing exactly when the hop-picking would begin. Finding work among the hops was a first-come-first-served business, and judging when to arrive was almost a fine art. Ideally, they aimed at arriving early enough to be among the first served, yet not so early as to have to wait too long without any means of support before the picking began.

Later in the century, things began to be better organized and it became possible for Londoners to book their places on the farms without leaving London until the night before picking was due to start. But at the beginning of the century that was not so. Then, they just packed up home, as people had already been doing for more than a century, and flocked into Kent, by road, afoot for the most part, with most of their belongings and such of their children as could not walk with them piled into wooden handcarts. The carts were not overburdened, and one would save the backs of several adults.

On the way they worked where and when they could by day, and by night they slept rough, often under the hedges or in open fields. When they arrived in the Kentish countryside, most were completely destitute.

Throughout Victorian times the demand for hops increased enormously. Beer consumption was high even before the duty on hops was abolished in 1862. When that happened, hop growing became very good business for the farmers of Kent.

There were several 'classes' of hoppers. First there were the gipsies. They moved around constantly, taking their caravan homes with them, wherever there was work for the whole family to do. They had regular 'beats' through Kent, as indeed they had through most of the counties, and hopping was just one of their regular seasonal occupations.

Then there were the 'travellers'—similar, outwardly, to the gipsies, yet distinct from them and, in the early nineteenth century, just fractionally higher up the social scale. They, too, travelled in their own house-carts, and often spent the whole of the harvest season doing casual farm work, sometimes for one farmer, sometimes going from one to another. Some of these were quite respectable, tidily dressed and quiet in their habits, asking only that they be paid for the work they did and trying to keep what we would call today a 'low profile'—keeping out of trouble as much as possible. Unfortunately, not all of them were like that, and it was the others that got the whole class a bad name with country folk.

Next came the 'foreigners'—anyone from outside the immediate locality of the hop gardens themselves. Mostly Londoners, but some Irish and even Welsh, with a spattering of people from other towns, usually outside the county. To some extent, even Kentish townspeople, from the Thames and Medway areas, perhaps, were 'foreigners' when they came to the hop gardens, although a fine distinction was drawn that allowed them to be a little less 'foreign' than the 'real foreigners'.

Finally, there were the home-dwellers: local country folk and villagers who worked on the hops all the year round from the first cutting away of the unwanted shoots from the old hills, through the training and tying to the picking. Later, the term 'home dwellers' came to be applied generally to almost all pickers who went to their own homes each night, as distinct from the Londoners who lived in the farmers' hopper-huts, and the gipsies with their own encampments.

First needs of the farmers for hop-picking were met by the home-dwellers, local farm-workers' wives and families and other local people, such as tradesmen's families. In Kent, the work force was usually about two hundred pickers for every acre of hops. They worked in companies under the direction of a binman/pole puller, who kept the pickers supplied with bins and moved the bins from stand to stand.

The binman was both overseer and spokesman; a sort of middle-man between the farmer and the pickers.

The earliest 'foreign' hop-picking labourers were mainly Irish harvesters and gipsies who sometimes lodged with the farmer and his family and sometimes were provided with accommodation in farm-yard barns and cattle sheds, in which case the farmer usually regarded it as his duty to provide straw for bedding and wood for fires, if very little else.

It was by no means uncommon for farmers to be burned out of house and home by labourers' farm-yard fires. Fires caused by candles left to burn among straw while their owners slept off an evening's drunken revelry, or by a pipe let fall into a bed of straw were often serious for the farmer and frequently fatal for the labourer. But, by and large, they were written off as natural risks one could not avoid while the need for casual labour existed.

But as the acreages of hops increased, so did the numbers of itinerant hoppers, and farmers could no longer adopt the same paternalistic relationship towards their casual workers. Instead of feeling obliged to house them, the farmers began to feel they were doing all that could be expected of them by paying the pickers wages. How the pickers spent the money, and how they lived when they were earning it, the farmers felt was no concern of theirs.

The fact that it cost the farmers very much less to pay wages than it would have cost them to provide accommodation for the growing hordes no doubt had something to do with this changing relationship too.

Picking usually started in early September and lasted three weeks or a month. Most of the pickers were women and children, although there were always some men with them. Most of them came from London and other towns, although at one time quite a lot of the pickers still came from Ireland and Wales.

Picking twelve or fourteen hours a day, a good season would see the hops cleared at the rate of about two acres a day. Early in the century, it was common for pickers to pick all day on Saturday, as well as every other day, but this meant that drying had to go on all through Sunday and this became unacceptable to Victorian morality so it became customary for picking to end at midday on Saturday. Then the dryer in the oast had Sunday free, to catch up on lost sleep and to attend church.

With the hop gardens empty of pickers on Saturday afternoon, hedgerows and clothes lines in and around the hop-pickers' camps were laden with the weekly wash. Oven fires were busy, pots brewing, cakes and pasties cooking, and the more conscientiously mothered children being bathed.

For the romantically inclined observer, it was a scene capable of being interpreted as charming and altogether delightful. It is, of course, the romantically inclined observers, rather than the pickers themselves, who have left us pen-pictures of the scenes in the hoppers' camps of this period. Except for fleeting hints here and there, we have very little idea of what the hoppers themselves thought about it.

The tally—that is the price per bushel paid to the pickers—was fixed each year by the growers according to their intelligence about demand. At first, it was very much a take it or leave it payment, but gradually that changed and there was a certain amount of negotiation, usually taking the form of protest by the pickers if the tally was fixed at a price they decided was unacceptably low.

From 1800 to 1810 the average tally was about a penny a bushel. Between 1840 and 1850 it varied a good deal according to the crop, from about three or four bushels a shilling to nine or ten. At the end of the century, tallies of about 2*d.* a bushel were fairly normal.

In the villages and on the farms, everything stopped for the hopping. Everyone who could be spared from other work made the daily pilgrimage out to the local hop gardens. Schools were always closed for the season, and the holidays were adjusted each year to suit the hop-picking period.

Shoes and pans went unmended as cobblers and tinkers turned hop-picker for the duration; there were no odd-jobbers during hopping, and the gentry had to adapt their needs and the needs of their homes for any sort of casual labour to take account of the hopping period. Church attendances always fell off more at hopping time than during any other harvesting.

But once hopping was over, of course, there was usually a boom in business, as the hoppers settled old debts and started new ones with the annual renewal of clothes and household items.

In 1835 Mr Ellis of Barming was perhaps the largest hop grower in Kent, with 500 acres under hops. He employed between three and four thousand hoppers each year. Most of those were Irish—many of them London-Irish—and they came to his farm regularly, year after year. It was his practice to supply his pickers with beer because, he said, if he did not then they drank water, and that meant a high risk of an outbreak of cholera in his pickers' camps. But in spite of the beer dole, he noted that most of the money his hoppers earned went straight away on spirits anyway.

Health was always a problem in the hopper camps. Hygiene was primitive to the point of being non-existent; food was often very poor—many of the hoppers virtually lived on potatoes, often stolen but sometimes provided by the farmers, and upon food brought to the hop gardens by London costermongers or other travelling salesmen. It was never prime quality and was often decidedly neither fresh nor clean, and certainly not wholesome. There was no piped water, of course, and what water there was either had to be carried quite long distances from the farm well or, more conveniently perhaps, from a nearby pond. There was no guarantee of purity, and a high risk of impurity, from either source.

Once in the hop gardens, the hoppers' lodgings were make-shift and primitive in the extreme. There was a good deal of overcrowding, as much because the Londoners were used to overcrowded living conditions and did not know how to avoid them as because the accommodation itself was inadequate.

Later reformers who studied the hoppers' conditions frequently referred to the indecency and immorality that was rife in these make-shift lodgings. The Reverend R. Shindler, in his book *Among the Hop-pickers in the Weald of Kent*, remarked:

> "Sometimes the overcrowding leads to much indecency. Some of the scenes frequent among the hop-pickers during the season are a grave scandal and a reproach to our boasted civilisation. It is as if the whole moral sewage of London had been suddenly disgorged upon the rural districts, spreading contagion wherever it flows."

Others writers confirmed his opinion. Visitors to the hopper camps were shocked by the bad language they heard, some of it directed at them, together with insults and abuse.

The Reverend J. Y. Stratton, who published his *Hops and Hop-pickers* in 1883, wrote:

> "So formidable and rampant was the vice commonly practised by the hop-pickers that many who longed to do them a kindness were afraid—the writer for one—to enter their lodgings or go among them."

Wife-swapping for the hopping period was commonplace, and so were hopper weddings, when couples were 'married' by all sorts of impromptu ceremonies in the sight of their comrades, and were regarded as perfectly free to part or return to more regular partners afterwards. One of the ceremonies commonly practised required the couple to jump together over a hop pole.

These hopper weddings were often supervised by the binmen, who 'allowed' men in their companies (the group of pickers for whom they were directly responsible as overseers) to take more than one wife for the duration of the hopping. There were all sorts of meaningless ceremonies, some of them coarse if not quite obscene, and the willingness of the women was sometimes in question to say the least. The practice could—and not infrequently did—result in considerable hardship and distress for one or both 'wives' who, when it was time to go home to London again, were 'divorced' with equal ease.

Some of the landowners were well aware of the kind of things that went on in the hoppers' lodgings. Some preferred not to know too much, and pretended not to be aware that anything irregular was happening. Some accepted such things as inevitable if not actually natural—as part of the 'foreign' way of life with which it was no business of theirs to interfere.

Some of the growers knew what went on and cared enough to try to stop it. One was so offended by his hoppers' immorality that he provided special lodgings with numbered and lettered accommodation which he allocated, some for married persons, some for single men, and some for single women. He no doubt felt he had solved the problem, but when he checked to find out how the arrangements were being received by the hoppers, he found they had ignored the numbers and the letters and were living, as they put it, 'all higgledy-piggledy'—and enjoying it very much more.

Nevertheless, there were hoppers who preferred a little decorum and it is on record that one party of eight or ten girls arranged with their binman to be padlocked inside their compartment in the hopper hut all night. In spite of the efforts of some of the young men nearby to wear down the girls' reserve with singing half the night, the girls simply joined in from where they were, apparently enjoying themselves quite wholeheartedly, but preserving standards of decency and morality that were comparatively uncommon among their kind.

From the middle of the century, concern about hop-pickers and the conditions in which they lived while they were in Kent built up, first from Christian individuals with a missionary inclination, and gradually bringing in local and national organizations, too. Bit by bit, the national conscience was awakened to the conditions of the people at the rough end of the beer boom.

The Reverend R. Shindler described the hop-pickers of the period very graphically. Most of the 'foreigners' (those from outside the immediate hop-growing area) were, he said, Londoners. Many of them he described as being in a state of extreme destitution, without either shoes or stockings, and often without a shirt or any other clothing than a ragged jacket or waistcoat and a tattered pair of trousers.

"One would think that water was exceedingly scarce, and soap an unknown thing, from the dirt and grime of weeks and perhaps months hiding the real colour of the skin."

For the most part, the only permanent structures available to accommodate the hoppers were the farmyard cart sheds, cow sheds, even vacated pigsties.

Some farmers saw no need to clean out the temporary lodgings before the hoppers moved into them, either. When they did clean them out, it was only a gesture, very often. It was considered by many farmers perfectly adequate for a few armfuls of old hop bine to be tossed in on top of the natural dirt or manure, and strewn with straw, to provide all that was necessary in the way of preparation for the temporary human tenants.

Hoppers thus housed could think themselves well served at that. Until well into the nineteenth century, many farmers thought they were fulfilling their obligations quite fully by making up little hoppers' lodges from hurdles laced with straw and fastened together with hazel bands. In fine late-summer weather, such shelters probably had an attractive, idyllic look about them and were probably even comfortable enough up to a point, but at the first drop of September rainfall or even heavy early morning autumn dew they would very quickly have become soggy, streaming invitations to acute discomfort and, almost inevitably, illness.

Later on, tents became common as temporary homes for the foreigners, but they were often not much better than the straw huts. Many of them were acquired by farmers from the Army which had condemned them as no longer suitable for soldiers, and they were long past their genuinely useful life as a means of keeping out the weather.

None of these temporary homes, of course, could be made in the least bit secure while the pickers were in the hop gardens all day and it was customary for a young boy to be left in charge of each one or, more often, of a group of them.

The provision of purpose-built hopper huts had to wait, on most farms, until the 1880s. Even then, and certainly before then, hoppers were expected to bring their own bedding with them if they were not content to lie directly on the straw that was usually provided. The only 'furniture' the farmers reckoned to provide was the straw for the floor, perhaps a few pegs or nails in the walls for clothes-hangers and, possibly, a tin pan or basin for washing.

The common feature of all hoppers' lodgings all through the nineteenth century was the straw that was almost always provided by the farmer. It helped to build the shelters, it was used for bedding, and it was used to litter the stamped earth floors. Some farmers felt obliged to renew the straw once or even more often during the hop-picking season. But if a season lasted, say, three weeks, many farmers thought that such munificence was quite unwarranted, and so the straw would remain, unchanged, throughout the whole three weeks, during which it was trodden into the ground with mud brought in on boots from the gardens, until it was barely recognizable as straw any more and was more like good quality compost.

Many of the hoppers arrived in Kent without the luxury of a change of clothing. Even if they left any behind—and many did not—such a refinement for a mere month-long stay would have seemed to most of them a wholly unnecessary extra burden on the long trek from London and back again afterwards.

If the hopping was a wet one, and the outside fires could not be lit or kept alight, there was nowhere they could dry their soaked clothes at the end of the day's work. The most sodden garments would hang up from lines strung across the room and their owners would snuggle down into their straw beds, huddled even closer than usual together and hope that the under-clothes they slept in would be a little less damp by morning. If the sun shone tomorrow, they would dry out while they worked at their bins.

Meantime, throughout the night, the dampness of the clothing and of the bodies and of the ground and of the atmosphere generally would move in with them and hasten the process of rotting the straw.

No wonder there were frequent outbreaks of epidemics of one kind and another. Unscrupulous itinerant traders sold the hoppers bad meat and bread; water was frequently both inadequate and unsafe to drink anyway; living conditions were appalling. Outbreaks of cholera were by no means uncommon and in 1849 one such outbreak killed forty hoppers at East Farleigh, including many children.

Childbirth was another hazard of the hop gardens—indeed, it was a hazard in the best possible conditions of the time. In the hop gardens the natural dangers of the period were multiplied many times. The Reverend J. J. Kendon, in one of his reports upon his missionary work among the hoppers, related the case of one woman who began labour in a hop garden. Her friends got her back to her lodge, where her first child was born. The child was immediately seized with fits, and

A Sunday-afternoon prayer meeting outside the barn in which hoppers of all ages and both sexes lodged during the hopping season in the 1870s

Only the fashions worn by the pickers date this hop-garden scene at Cranbrook in 1906

Measuring out a bin at Beltring in 1932

1933: The Salvation Army provided crèches for the very young children of the hoppers at many farms

died two days later. Such incidents were almost commonplace.

All the calamities that befell the hoppers were not the result of natural causes. In the first half of the century a party of thirty hop-pickers—gipsies and Irish labourers—were returning to their camp at the end of the day's picking behind another cart-load of home pickers, and had to cross Hart Lake Bridge near Hadlow. The home pickers crossed safely enough but the horse pulling the second cart took fright and went through the bridge rail into the river, taking the cart and its human cargo with it. Recent rains had swollen the river and thirty people drowned that evening. They were all buried together in Hadlow churchyard, where a monument still commemorates the tragedy.

The Reverend R. Shindler saw fit to find a moral for readers of a tract he wrote after the event, in which he pointed out:

> "There is something mysterious (about the incident), for those 30 persons were all gipsies or Irish and not only had a party of home-dwellers been conveyed across the bridge in safety, a short time before, but certain trivial circumstances had prevented several residents uniting with the party that met with so melancholy an end.
>
> "Someone may say: 'This fact explains all; they were notorious sinners, I doubt not, and God has punished them for their sins with a dreadfully sudden death'."

But the parson's charity forbade him to leave it there. He went on to suppose that in fact the doomed party may have been no worse than some of the home-dwellers.

FIVE

Nineteenth Century—Second Half

Fine Betsy the bunter from London,
From out of St Giles's did prance,
Young Roger the black chimney sweeper,
The Grinder and his bonny lass.

They had all been drinking of gin
Till most of their money was spent,
Then agreed to ramble together,
To go a hop-picking in Kent.

Next morning they all met together,
The devil a farthing they'd got,
The grinder he pawned his old barrow,
Poor Nelly she pawned her new smock.

They set out as jovial as could be,
To Maidstone it was their intent
To meet with their jolly companions
To go a hop-picking in Kent.

Town Malling it was the first place,
The farmer he gave them employ,
They slept all together in a barn,
Like so many hogs in a sty.

Next morning they all met together,
Their tea to get under the hedge,
Their kettle was flung with three sticks
And they all around it did wedge.

When breakfast was over and done,
They all to the hop yard did trip,
Some helped to pull up the poles
And others the hops they did pick.

They spent all the day with great mirth
Nor thinking they done any harm,
At night when the day's work was done
Like hogs they run home to the barn.

When hop-picking was over and done,
The farmer he gave them a treat,
They danced like rams in a halter,
When each jolly couple did meet.

Young Roger the black chimney sweeper,
The Grinder and his bonny lass,
And Betsy they made her so drunk
She was forced to lie down on the grass.

The farmer he finds them in milk,
Ground ivy it serves them for tea,
They live all together in a barn
No landlord or taxes they pay.

Their life is a compound of ease,
They never do seem to relent,
And this is the life of those jiggers
When they go hop-picking in Kent.

Victorian ballad

Throughout the whole of the nineteenth century, and certainly during Victorian times, the double standards that characterized social conditions everywhere made it possible for people who would never have considered for a moment actually going hop-picking themselves to see the lives of those who did as picturesque and enviable.

A good example of this was provided by Chambers' Book of Days in 1869, in which part of a long descriptive introduction to the month of October extolled the delights of hop-picking.

"There is still one out-of-door scene beautiful to look at and pleasant to walk through, and that is hop-picking—the last ingathering of autumn that finds employment for the poor; nor are there many prettier English pictures to be seen than a well-managed hop plantation.

"The smell of the hop is very delightful, so different from that of new hay and hawthorn-buds, yet quite as refreshing. . . .

"Merry people, too, are the hop-pickers, whether at their work, or when going to or returning from the hop-plantations. The little huts they run up to sleep in, their places of cooking, washing, and other domestic contrivances, tell that they belong to the race who have heralded the way into many a wilderness, lived there, and founded colonies, that are now springing up as great nations.

"We see them travelling to the hop-grounds with baby on back, and leading children by the hand, carrying nearly everything their humble home contained.

"We look on and wonder how those tiny bare feet will ever tramp so far, yet while turning the head and watching then, we see them go pit-pat over the ground, three or four steps to the one or two longer strides of their parents, caring no more for the gravel than if they were shod with iron, and we are astonished to see what a way they have gone while we have been watching. Sometimes in the hop-grounds, we have seen a cradle with the baby asleep in it, swinging between the tall hop-bines, and thought what a pretty picture it would make if well painted!"

Well, so it would have, no doubt. Indeed, so it did, for it was painted and sketched. But the pictures hid more than they revealed of what the life of the hop-pickers themselves was really like in those days. Almost anything can be made to sound idyllic if not too many questions are asked.

Of course, hop gardens do have a beauty of their own. The Kent-born botanist Anne Pratt may well have been reporting the literal truth when she wrote:

"Travellers who have beheld, in other lands, the various scenes of culture—the olive grounds of Spain or Syria, the vineyards of Italy, the cotton plantations of India or the rosefields of the East—have generally agreed that not one of them all equals in beauty our English hop gardens."

And Charles Dickens may have allowed David Copperfield to speak for him when he had the young man say:

"I thought it all extremely beautiful and made up my mind to sleep among the hops that night; imagining some cheerful companionship in the long perspective of poles, with the graceful leaves twining round them."

A hop garden is very beautiful. And there is no doubt that, even with the conditions that prevailed at the time that little ditty and that almost lyrical account in the *Book of Days* were written, hopping down in Kent did have a great deal of attraction for many of London's East Enders, too.

That conditions some of them found when they reached Kent were almost unbelievably squalid and sordid by modern standards did not worry Victorian hoppers as much as they might have done if they had ever experienced modern standards, but, of course, they had not.

Many of them spent their everyday lives in damp, dark, airless slums—domestic as well as industrial—where drainage and sewerage were very primitive indeed; where a pump or possibly a stand-pipe served dozens of families with their daily water; and where rats and mice were at least as numerous as people and lesser vermin very much more so.

The family home for London's poor at this time was, at best, a small, dark room, usually in the garret or cellar of some crumbling tenement. Some lived permanently in the streets, in dark warrens of lanes and alleyways, where open ditches served for sewers and drains. Often, one bed was a whole family's only furniture and it would sleep eight or so people of all ages and both sexes. They wore the same rags day and night, and if they had food to eat it was more often than not unfit for human consumption. At a time when the poverty line was drawn at about fifty pounds a year, many had no regular income at all, and lived by begging, stealing and very occasional casual labour.

There was Poor Law provision for the completely destitute, but its administration was severely corrupt and always inadequate. Poverty was generally regarded by the people who paid for its relief as a self-inflicted condition, and Poor Law provision was deliberately made unattractive to deter people from taking advantage of it.

There was also a fierce independence of spirit among the poor that made any acceptance of charity abhorrent to all but a few. People tended to be more concerned for their personal 'liberty' than prepared to be 'bullied into health'.

It was common for lodging houses to provide accommodation for mixed sexes and inmates often shared a bed, men and women all together, and used a pail in the middle of the room as a latrine.

The conditions to which London's poor were accustomed in their own territory, so to speak, need to be borne in mind when the conditions to which they submitted in Kent are described.

Many of the Kent hoppers came from the foul 'rookeries' of

nineteenth-century Whitechapel and Southwark, and as Charles Wood, writing in *Argosy* in December 1883 commented:

> "Imagine for a moment the change from a stifling court and room, where the sun never shines, where pure air is never breathed, where cleanliness is unknown, to sudden transportation to one of these hop-gardens."

Well, now, put like that. . . .

The fact was that if these people came to Kent to living conditions no better but really not all that much worse than they were used to, with at least in September a good chance that the stifling late summer heat of the city would be exchanged for the aromatic sunny warmth of the autumn countryside, they could afford to regard the expedition as a very fair exchange and something of a holiday.

And that they did.

The second half of the nineteenth century was marked, in the Kentish hop gardens, by a particular cause and effect sequence that characterized the whole period. First, it was a time when demand for beer and, therefore, for hops, was still on the increase. In order to meet this demand, it was necessary to speed up the old methods of drying hops and getting them to the markets and so to the brewers. As it became possible to dry more hops, it became necessary to pick hops faster, in order to keep the kilns loaded to capacity. This meant the numbers of pickers had to be increased and led to worse conditions, as more people were crowded into the same never spacious accommodation. This gradually led an awakening social conscience to demand improvements.

These demands came first from churchmen. Although their first concern was for the spiritual wellbeing and moral comfort of these largely godless Londoners, their pursuit of that goal brought them face to face—as the middle classes had never been before—with the realities of what were so often pictured to them as sylvan idylls of life in and around the hop gardens.

The increase in the numbers of hoppers arriving in Kent was quite sudden, and many farmers saw no reason to spend any part of the extra income they derived from the new market trends on improving the lot of the pickers. After all, they reasoned, it was only for one month in twelve that hop-pickers' accommodation was needed at all.

Besides, the Londoners were far from fastidious and although they would complain bitterly and even protest actively if they thought they were being short-paid, they were not so ready to object to being huddled together into dark, damp sheds.

Even when the farmers were prompted by the reformers to wonder if their hop-pickers needed such refinements as latrines, they were very likely to decide that any such thing was entirely unnecessary. The pickers themselves certainly never thought them worth demanding. Indeed, many of them would have thought latrines a rather effete amenity in the open countryside.

Early in the century, disputes over pay or (less often) conditions had led to strikes and riots, and although they were usually settled by the intervention of the local police, sometimes it had been necessary to call in reinforcements from nearby towns. On rare occasions the military had to be called in.

As the years went by there was less of this, partly because of the gradual improvement in conditions (though not necessarily in pay) that took place. Even so, growers were still, quite often literally, at the mercy of their pickers, led by semi-professional bully-boys and militants. Farmers were liable to have their windows broken, and it was not unknown for pickers to threaten or even try to set fire to farmhouses in their efforts at intimidation and extortion.

Not all the bad feeling was directed against the farmers, either. Local pickers sometimes tended to side with local growers and then, with the Londoners banding together against the locals, mob rioting would break out. Until the forces of law and order intervened, it usually went badly for the locals, who were quite simply terrorized by their visitors.

The foreigners never had been much better than just tolerated in the rural districts. While their numbers were modest, the locals were a reasonable match for them when any trouble did occur. Now that the Londoners were arriving in very considerable numbers indeed, however, the tables were turned. Ill-feeling between the locals and the visitors was common and was sometimes demonstrated with no other instigation than mutual animosity.

Nor was it only villagers who fell foul of the quick-tempered hoppers. In 1863, the Mayor of Maidstone accused an Irish hop-picker of assaulting his wife as she got off a train. The

Irishman turned his belligerence upon the Mayor who was soundly beaten up on the spot.

There might well have been earlier moves to improve the lot of the hoppers if would-be reformers had been less afraid of the 'foreigners'. Gradually, this very understandable reticence was overcome and efforts to help the hoppers began.

One of the first of those whose reforming zeal found expression among the hop-pickers in the Weald of Kent, and whose work there was described so enthusiastically by the Reverend R. Shindler, was the Reverend J. J. Kendon, who made his headquarters at Curtisden Green near Goudhurst in the 1860s. By 1889, his mission was employing thirteen missionaries in the parishes of Lamberhurst, Yalding, Curtisden Green, Goudhurst, Marden and Horsmonden.

In his first report on his missionary work, among the hoppers, Mr Kendon wrote:

> "Here may be seen the dregs of many of our larger towns, and of London. They sleep in stables, lodges, barns, hopper-houses, straw huts, etc., almost like the cattle of the field. To mingle among these poor creatures, to see their habits and hear their language; to witness the awful lengths in wickedness to which they go, makes it seem almost impossible that we can be living in England in the later part of the nineteenth century."

Kendon reported meeting two little bare-footed boys who were not related, but were both from London. He spoke to one of them and was told that the other had been brought hopping by a man and woman who had asked his mother to let him come with them. Once in Kent, fifty miles from home, the man and the woman had beaten the boy, given him nothing to eat, and then turned him off because he couldn't pick enough hops for them. So now, said the boy, the waif was picking with his family, and they were feeding and bedding him with them.

The Reverend Shindler, recalling his first journey to preach to the hoppers at Brenchley, wrote:

> "I caught them just as they were. The sight impressed me as exceedingly picturesque. In a large field a number of house carts were standing under the shade of some spreading oaks. No-one seemed to have put on Sunday attire. The people were sitting or standing, lying or lounging about in all attitudes . . . the men

smoking, the women talking, or trying to hush their babies to sleep. . . .

"The behaviour of the people was decorous."

There was no doubt that hopping down in Kent did attract some very unsavoury characters, even for those times. But the hoppers, then as later, were a very mixed bunch as Mr Shindler acknowledged when he commented:

> "He who enters a hop-garden with the idea that all the persons he sees are thoughtless, heedless, hardened and Godless people labours under a great and serious mistake."

One of his particular concerns was that, apart from the month of hopping, when the hopping missions were active, the gipsies were without any contact with church or religion at all, and it was said of hoppers generally that they heard more of the gospel during hop-picking, thanks to the missions, than they did all the rest of the year.

Of course, they were a captive audience for the evangelists, but although the missionaries often began their work with the idea of saving souls, many of them quickly recognized that the most urgent need was to help to keep body and soul together. One of the ways Mr Kendon popularized his mission at Curtisden Green was by starting a custom of giving free teas. His missionaries would go into the hop gardens, where they talked to people at their bins, and handed out tickets to as many as would accept them, for a free tea.

Some of the hoppers would not take the tickets, but many did for whatever else they lacked, they always had good appetites and the teas became very popular indeed with the children. Vast quantities of tea and buns were consumed, and after the young hoppers' bellies were filled, the missionaries did what they could to fill whatever voids there might have been in their guests' souls with a prayer-meeting and hymn singing before the hoppers returned to their huts.

In the preface to his book *Among the Hop-pickers in the Weald of Kent*, the Reverend R. Shindler wrote:

> "The Gospel is distinctly and emphatically the weapon of the warfare of the Agents of the Hop-pickers' Mission; it is the message they bear to the sin-sick, the sorrowful, the guilty and the perishing among whom they labour. . . ."

Mr Shindler did not have a very high opinion of most of the people who came hopping down in Kent in his day.

The Reverend J. Y. Stratton was one of those who helped to publicize the work that went on to try to improve the lot of the hoppers in Kent. In 1865 he was drawing public attention to the seamier side of life on the Kentish farms at hop-picking time. In that year, he wrote an article which was published in *Leisure Hour*, in which he referred to a number of plans for improvement of hop-pickers' conditions.

These included the appointment of an inspector of hopper huts; a requirement that the huts should be wind- and rain-proof; that adequate latrines should be provided; and that an inspector should be appointed by the Board of Guardians of each district to inspect food and water supplied to the hoppers. He wrote of the need for sufficient floor space for each person in the huts, and for screens between the beds; and for the huts themselves to be clean when the hoppers moved into them.

The ideas were not his own; he was only publicizing a need that was already being recognized by the Kent growers themselves. A meeting of growers and landowners had already been held in Maidstone, at which it was decided that these very suggestions were acceptable and should be implemented.

In fact, it was a group of principal land-owners and tenant hop farmers who met together and formed the Society for the Employment and Improved Lodgings of Hop-pickers. The meeting was held on October 25, 1866, and it made a number of recommendations about standards of hoppers' lodgings. In 1867, the Society added a rider which acknowledged that "while all reasonable care should be taken for the encouragement of morality and good order among the hop-pickers, great caution was needed in dealing with the arrangements in the lodgings which the occupants prefer to make for themselves".

In other words, while the farmers recognized their responsibility for encouraging what they would regard as the better behaviour of their pickers, they also recognized that it would have been unwise for them to try to interfere too much.

The Society for the Employment and Improved Lodgings of Hop-pickers agreed that for purposes of floor space, two children under twelve years old should equal one adult. They also agreed that a screen of metal or wood, 6 feet by 4 feet, should be fixed between each bed. Each room in a hoppers' hut, they said, should have a window in the roof, 18 inches by 14 inches, and the doors should have a bolt inside and a padlock outside. There were to be shelves and cupboards where possible, and clothes pegs were to be provided. Iron sconces were to be driven into the walls to hold candles, and lanterns were to be provided.

Disciplinary control was provided for by an agreement to make binmen responsible for all loss or damage to the huts, and provided equipment, which should include one oven for every twelve companies of pickers. Instead of straw litter for beds, hoppers were to be provided with hop pokes filled with straw to make mattresses.

It was decided that huts of wood or straw were too great a fire risk, and that hopper huts should in future be brick built. Where tents continued to be provided, they must be weather-proof in future, and dressed with tar and cod oil to keep them that way.

The standards set by the Society for future hopper huts were ahead of their time. In fact, they were too far ahead of their time, and although the members, by and large, honoured their approval of the improvements, there were a great many farmers who did not—for various reasons—go the whole hog. Hopper huts of wooden construction, for instance, were to outlast the nineteenth century by many decades.

Perhaps one of the most important of all the reforms instituted by the growers in the interests of the better management of their hop gardens and the pickers who came to them, was the appointment in 1867 of London hiring agents, operating in collaboration with a corresponding agent based in Maidstone.

Until this time, there had been no organization of pickers. They just came to Kent when they thought picking was due to begin, and hired themselves out to any farmers who needed them on a first-come-first-served basis. This was satisfactory enough, for the farmers, at least, while numbers were manageable but as they increased, the problems of large numbers of unemployed vagrants wandering hopefully about the rural districts became very considerable indeed. Some sort of control became essential.

Until the 1860s in London or the suburbs, a man would agree with hirers to bring down one or more bin companies of six or ten people plus their children. Sometimes the gangers took all the money, and paid their companies as little as they could get away with. The gangers arranged for their own companies to be fed and this opened up more opportunities for racketeering by making special arrangements with suppliers—special arrangements that gave the gangers the greatest profit and, usually, left the pickers very poorly provided for indeed.

Even if the pickers were paid separately, it was usual for them to pay their binmen for 'looking after' them, and this could cost them as much as five shillings a head out of their hard-earned hopping money.

If, as could happen, a hop crop succumbed to one of those attacks of weather or disease that caused them to 'go off' (wilt) suddenly, then the pickers would arrive and find they had no work. When that happened, it was a tragedy for the grower, who might lose a great deal of money from such a mishap; but it was a very great tragedy for the pickers, too, who often had no alternative but to turn back and beg their way home again, on foot, with no money with which to buy food or lodgings on the way.

Sometimes, rather than go through the London hirers, the growers would simply hire all the pickers they needed from the local union-houses and, knowing that this was done, the Londoners came to Kent, destitute, and threw themselves upon the mercy of the union-house keepers, enormously increasing the numbers to be accommodated until the farmers took as many as they needed.

In 1865, more than one thousand, one hundred people journeying from London to the Kent hop gardens slept on doorsteps in the streets of Gravesend alone. Thousands more did the same in other towns of North and Mid-Kent. As they got nearer their goals, they slept rough under hedges or in open fields. In September 1869, the *Maidstone Journal* reported that in eight days 618 men, 817 women and 1,800 children (3,235 people in all) had applied to Maidstone's relieving officer for lodgings during the night and for food during the day. Coxheath Union, four miles from Maidstone, was, at the same time, taking an average of 600–700 people every night.

On the other side of the North Downs, the Medway Union at Chatham took in 2,000 casual lodgers during one week—and so disorderly were they that the master there had to call in the police to subdue them.

In 1876, the Chief Constable of Kent recorded 100,000 immigrants into the county. They were an enormous burden on the county services, for they had no means of their own at all. What they could not beg, they stole. Outbreaks of disease, which were common, were liable to assume epidemic proportions almost overnight.

The introduction of the hiring agents did not solve these vexing problems overnight. Indeed, the control they exercised in theory was for some time a good deal less than satisfactory in practice.

The object was for the agents to make up lists of people who wanted to go hop-picking. The lists could be started almost any time during the year, but as picking time drew near, the Maidstone agent was able to tell his London counterparts which farmers wanted how many pickers, and the London men formed up their companies of pickers and binmen from their lists in numbers intended to meet demand.

Each picker thus recruited was given a Hoppers' Letter, which had a registration number, the picker's destination, train times and the name of his employer. The idea was that the picker should take the letter with him as his introduction to the farmer and, at the end of the picking, the farmer would endorse the letter with his comments about the suitability of the picker for similar employment another year. The endorsed letter was delivered up to the London agent who could then draw up a black list of pickers who were not to be given a place on his list next year.

The advantages of the system were not immediately recognized by the pickers, many of whom preferred still to come on the off-chance. There was, too, a good deal of abuse of the system, both by pickers and agents at first. But gradually the rogue agents were weeded out and the Hoppers' Letters became practically essential as a passport to work in the Kentish hop gardens.

With tighter control over the numbers arriving in the county, it became easier to improve the conditions to which they came, as well. When, in 1868, the Hon. E. Stanhope, MP, personally inspected accommodation for hoppers in Kent, he reported: "More than half the pickers are now decently lodged and the improvement which takes place in every successive year is very marked."

Well, so it was. But it still left much to be desired.

In 1870, the *Maidstone Telegraph* reported an inquest held at the King's Head at Bloomsbury, London, on one Margaret Goderich, aged forty, who was said to have left her home in Long Acre with her husband to go hopping in Kent, and found work at a farm at Marden. She slept in a tent made of a few pieces of canvas that left the occupants almost wholly exposed to the weather. It was said in evidence that it was a very wet hopping, and people had to go to work drenched to the skin each morning. As a result, the deceased's lungs became inflamed and she was unable to work. In that condition, she had to find her own way back to London as best she could, leaving Marden on Friday and arriving home on Monday night. When she finally reached her home, she went to bed and died the next day. Just before she died, a Dr Bennett was called to attend her and he told the inquest that he had other patients in a similar condition as a result of going hopping.

The Coroner, a Dr Lankester, said he considered the system under which hopping was carried on required 'amendment'. But that didn't help Mrs Goderich and Dr Bennett's other patients in a (one supposes, not quite) similar condition. The jury returned a verdict that she died from inflammation of the lungs, caught from sleeping in a wet tent while hopping at Marden.

When the Local Government Board was created in 1872, there was no legislation covering hopper huts, and the accommodation provided for hoppers was still far below what many people by this time regarded as acceptable. Over-crowding was still, in many cases, extreme and the cleanliness of the accommodation hoppers were offered when they arrived at the farms often left very much to be desired. Many of the huts themselves were still serious fire risks; separate lavatories for men and women were quite uncommon; and wash-houses were luxuries indeed.

The recommendations of successive reformers first achieved the weight of law as a result of model by-laws which, however, the local authorities were empowered—not compelled—to adopt. Some did adopt them, some did not. Some of those who did were less than scrupulous in enforcing them, and some fell into general disuse subsequently.

The first by-laws regularizing hoppers' accommodation and

conditions were adopted at Bromley (then firmly inside the administrative as well as the geographical county of Kent), under the Sanitary Acts Amendment Act 1874 (as again amended in 1875), which provided that any local authority might make by-laws for securing "the decent lodgings and accommodation of persons engaged in hop-picking within the district of such authority".

A model code of by-laws was provided by the Local Government Board, with provision for a five pound penalty for offences against the by-laws, and a further ten shillings a day penalty while the offence continued.

After Bromley had led the way by adopting the model by-laws, Cranbrook, Tenterden, Milton (near Sittingbourne) and North Aylesford followed. Maidstone was among the first of the local authorities to adopt by-laws covering hoppers' conditions, too, although there the by-laws that were made were not those suggested in the Board's model.

The 1875 model by-laws had, in fact, been already largely overtaken by improvements introduced in the Maidstone district which was why the Maidstone councillors did not adopt all of those in the model. Unfortunately, though, once Maidstone set this precedent of making its own—different—by-laws, other districts followed suit, not all of them with such good reason. Hollingbourne, for instance, was one of these and the result was that here and elsewhere the by-laws' intentions were so eroded that the hoppers were left little better off in many districts, certainly in law, and often in fact, than they had been before.

The September 1878 issue of the Kent and Sussex *Times* printed a letter, in rhyme, from a hopper who complained about the way the hoppers were treated. His letter included the verses:

> "The farm that I work on has one hundred baskets or more, And the hops that we pick are dreadfully poor. To cheat us of our measure the tallyman's bent; That's how we are treated when hopping in Kent. The huts that we sleep in, God knows, are well named: They are 'huts', not dwellings, and let in the rain. They were not made for comfort. So pray be content And never grumble at trifles when hopping in Kent."

By 1883, of thirteen districts comprising 233 parishes, six had framed by-laws on the Local Government Board model, and adopted the main provisions of the model. The rest had their own by-laws which left out provisions for such things as cooking sheds, screens between beds, and latrines.

Many of the reforms that gradually took place came as a result of the hoppers themselves banding together to demand improvements. However, the improvements achieved in this way tended to be short-lived. Concessions won one season usually had to be fought for all over again the next.

In fairness, many growers were not at all satisfied with the general standard of conditions for hoppers in the county as a whole, and some of the bigger land-owners were foremost among the reformers seeking improvements. It was they who, as loudly as anyone, urged that the permissive local authority powers ought to be made compulsory, and that a uniform code of by-laws should be introduced.

The growers generally approved and welcomed the hiring agent system. Five agents had been appointed by the Society for the Employment and Improved Lodging of Hop-pickers in 1867, at a retainer of fifty shillings a year plus one and sixpence for every bin's company of ten pickers, or a proportionate sum where pickers were engaged without binmen. The Society charged growers threepence a picker for this service.

Generally, too, the pickers approved and supported the system. Binmen, of course, did not. Hitherto, it had been customary for them to gather their companies of pickers and take them to the hop gardens where they levied their own toll for looking after members of their companies. The new agency system took away a lot of the power of the binmen over the pickers, and left them without many of the 'perks' they had been used to.

Nevertheless, on the whole the system worked well. Registers were kept, and people who wanted to stay on those registers knew they had to behave themselves while they were in Kent. It was this hiring agency system that introduced a period that was to last as long as the old-style hopping itself lasted, when families came to the same farms year after year—indeed, generation after generation—and became as well known to the farmers as their year-round local families.

Although the new system got rid of some of the old rackets there were, inevitably, new ones. When pickers struck for better pay or conditions in mid-hopping, growers broke the strike simply by sacking the strikers and then using the agents to provide replacements from among the names on their registers. There were always more people wanting to come hopping than the growers had need of.

Railwaymen, too, were known to have been bribed to misdirect illiterate holders of hoppers' letters, so that other pickers could fill the vacancies on the farms.

The hoppers' letter issued by the London agents included details of who the holder would be working for, and some applicants took letters 'for all the family' and then sold those he didn't need. They fetched twopence each from neighbours and friends who did not properly understand the system. Actually, there was no need for anyone to buy a black market letter in this way, unless they had already been 'black listed' by the agents. Anyone else could get on to the registers and qualify for his own hoppers' letter free.

Gradually, changes did come to the hop gardens. By the last decade of the nineteenth century, when there were about 250,000 hoppers coming to Kent every season, some 70,000 or 80,000 of them from London, a Lamberhurst landowner, Mr John Noakes, was able to report to a Parliamentary Select Committee on the Hop Industry: "It is very offensive to be near them when they come, but when they go away, with a little money in their pockets, after a month in the country, they are clean, respectable sort of individuals."

Even so, the provision of purpose-built hopper huts had to wait, on most farms, until the 1880s and later. Local sanitary authorities, which had powers to require farmers to build proper hopper huts, generally did not do so, preferring to leave it to the individual growers to provide the best accommodation they could. Where hopper huts were built, they were most often one long, low-pitched wooden shed, divided up inside with sacking partitions to provide ten or more compartments or rooms. Some—and, gradually, most—had wooden partitions, usually only half-way up the height of the building, and often not solid, but of spaced boards or lattice-work. They were just about enough to keep families separated from each other, providing a very minimum of privacy for each family, and encouraged a freer passage of air circulating through the building than would otherwise have been provided.

But many hoppers still spent their hopping in make-shift

sheds. The Reverend J. Y. Stratton, writing in 1885, described how he visited a cart shed that was doing service as a temporary home for some of the Londoners. His visit took place in the early evening but, already by then, the shed was almost dark inside.

There were no internal divisions. In one corner slept four young men. Next to them was a group of married people and, in the opposite corner, the young girls of the group.

We may suppose that Mr Stratton had good evidence, and was not relying upon what he might have regarded as the inevitable consequence of this sort of togetherness, when he wrote:

> "On the first Saturday night, with the wages gone in food and drink, good intentions were abandoned, and long before the close of picking the shed was a den of iniquity."

It is possible to detect a note of tolerant despair in the writing of the Reverend Stratton when he noted that one of the great difficulties was to maintain a "proper standard of ventilation" in the hoppers' lodgings. The good vicar was strong on ventilation, and certainly in the sort of crowded conditions the hoppers experienced in the Kentish hop gardens, well-ventilated premises must have been almost essential for the preservation of even moderately good conditions.

But: "Some ingenuity is displayed in ventilating modern buildings", he reported, adding however that ingenuity was necessary because the London pickers abhorred ventilation, and exercised a good deal of ingenuity of their own in making sure they had as little as possible. They hung sacks over the open-work partition walls, and they used straw-laced hurdles, bundles of twigs and branches, or indeed anything they could beg, borrow or steal to complete the floor-to-roof partitioning.

Each compartment in a hopper-house usually had its own door and this, too, generally had a sizeable gap between the top of the door and the frame. That would be stuffed up with rags or paper when the occupants settled down for the night. The Londoners were not used to fresh country air, and were not specially anxious to get used to it.

At Bridge, near Canterbury, the local council made regulations requiring that temporary huts should be provided with a dry foundation and designed to give 250 cubic feet of space for each adult as a minimum allowance. The same regulations required there to be separate closet accommodation for each sex in the most secluded spot available, and plenty of pure water for cooking and general domestic purposes.

The council also warned hoppers:

> "As it has hitherto been a common practice for hop-pickers to bring along with their families some amongst them who have recently recovered from communicable diseases, notice is hereby given that any person suffering from any dangerous infectious disorder, who wilfully exposes himself in any public place, or any person in charge of one so suffering, or any person who without previous infection exposes any bedding, clothing or other things which have been exposed to infection from such disorder, will on conviction be liable to a penalty of £5."

In 1885, when those regulations applied, the threat of infection brought to Kent by the hoppers was very real, and a matter of considerable concern to local authorities.

By the turn of the century, though, the bad old days of hopping down in Kent were passing rapidly, merging into the good old days that many hoppers looked back on with real pleasure right to the end.

In August 1887, an article in *The Echo* was able to look back on former times when "the picking of hops brought with it into the peaceful villages, where the gardens are situated, a terrible increase of crime". The writer declared that that was all changed. The Sanitary Inspector, the Excise Officer, the constable and "last, but by no means least, the self-denying band of Christian workers" had so changed things that crime was scarcely heard of, drunkenness was infrequent, and the magistrates of Kent and Sussex found their work but little increased by the sudden influx of these vagrant thousands.

Where the writer got his information from is not known. Certainly there were plenty of authorities in Kent who would have disputed the implication that all in the hop gardens was now lovely, but changes there certainly had been, as much in the people as in the conditions they experienced.

SIX

Twentieth Century—1900–40

The early sun serenely shines upon the Kentish hills,
Gilding the corn, the hops and bines, the rivulets and rills.
'Midst the perfumed fields a stranger stops, for love had just been there,
His trap was laid 'neath the amber hops, 'twas a maiden sweet and fair.
And there on that autumn morning Cupid's bow was bent
And straight without a warning the darts to their hearts he sent.
He gave them his old sweet greeting, then off with a laugh he went,
Leaving these two hearts beating with love midst the hops of Kent.
Then as his bride he took the maid forever in his care,
Away from the wood, the fields and all, the meads with their sweet scent,
Though left the scene, they oft recall, their bliss midst the hops of Kent.

Victorian version of an old song passed down through the family of Miss W. D. M. Constable of Stoneings Farm, Knockholt for 200 years

The twentieth century will undoubtedly be remembered for many things, not least for the rapidity of change that it brought to almost every aspect of life.

Farming changed more slowly than many other features of life—it often does in times of rapid change; there is no hurrying the seasons—but even on the farms it was the twentieth century that brought to an end hop-picking traditions that had survived for 300 years.

It would have been a far-sighted farmer or hop-picker, though, who would have forecast at the beginning of the century that, by the middle of it, hop-picking as it was then would have ended completely.

With seasonal ups and downs, the hop-growing boom was still on, and there were still very large numbers of 'foreigners' flocking into Kent every year, coming for their annual hopping down in Kent. In 1908, for example, the County Medical Officer, Dr William Howarth, estimated that there were about 74,748 foreign pickers scattered through the county. That figure excluded about 5,600 gipsies and an uncertain number of home pickers (although there were said to be 12,000 home pickers in the Tonbridge area alone).

Most of these visitors came to Maidstone (27,500), Malling (10,500) and Tonbridge (10,200). Fewer found work in the Ashford (2,500), Bridge, near Canterbury (3,314), Cranbrook (6,231), Eastry (1,200) and Faversham (about 9,000) districts.

The nineteenth-century reformers had almost won their fight by 1900. There was still plenty of room for improvement in hoppers' conditions on the farms but, at least, the need for these improvements was firmly established by now. The hoppers were, more generally than they had been fifty years before, recognized as people, and as important to the well-being of the hop industry as the growers and the factors and everyone else who contributed to its overall prosperity.

The conditions of hoppers in Kent were reviewed for the Local Government Board in 1907 and again in 1914, and it was generally agreed that there was no serious overcrowding any more, and most farmers were by now providing reasonable facilities for the maintenance of health and hygiene in the hop gardens.

It does not follow, of course, that we would, today, accept 1907 or 1914 assessments of what were 'serious' overcrowding, or 'reasonable' facilities. In fact, some authorities at the time and even long after that still found fault with both the accommodation and the facilities provided for the hoppers. Some farmers still baulked at providing anything but quite crude cookhouses and wash-rooms or even lavatories for their hoppers who, they argued, used them only for a few weeks each year, and sometimes ill-used them even then.

Ordinary farm buildings, including barns, stables and cart sheds were still being used. Some farms were still providing bell tents, usually second-hand ex-Army equipment which were all too often very defective. However, a large number of special wooden hopper huts had been provided as well by now, although not all of them were properly weather-proof and the absence of paths to them or concrete areas around them meant that in wet weather, such as attended the season of 1908, for instance, the whole area was a mess of squelch and mud.

The difference between 1908 and, say 1808 or even 1858, was that instead of being regarded as inevitable and natural, such conditions were now officially condemned. The Public Health Act of 1875 had empowered any local authority "if it saw fit" to make by-laws securing the decent lodging and accommodation of persons engaged in hop-picking. The model by-laws provided by the Act required each adult to have sixteen square feet of floor space (with two children under the age of ten assumed to be equivalent to one adult for floor space purposes). Whatever habitation was provided for them was to be clean, dry, weather-proof, ventilated and light. Screens or partitions were to be used to secure the privacy where adults of different sexes were accommodated and there were to be separate cooking places for every fifteen persons. Good water for drinking, cooking and washing had to be provided, and there had to be adequate privies for each sex.

By 1906 many of the rural district councils had adopted by-laws based on these models, but not all the farmers in their districts were complying with them, and enforcement proceedings were not common. It was still very much a matter of the lengths to which the farmers themselves were prepared to go to make their hoppers' stay in Kent a comfortable one.

Before 1914, a brick hut of twelve rooms, each room about 10 feet by 8 feet 6 inches by 6 feet 6 inches, suitable for twenty-four adults or fewer if they had children in with them, cost about seventy-five pounds to provide. Many farmers needed half a dozen or a dozen such blocks of hoppers' accommodation, plus separate men's and women's lavatories, one or more wash-houses and a cookhouse. It all added up to real money that only earned any kind of return at all for about three or four weeks of each year.

Nevertheless, most farms did have hopper huts by the beginning of the First World War, even though many of them were still the old wooden kind with corrugated iron roofs.

In 1913 the County Medical Officer, Alfred Greenwood, reported in his Annual Report that the Local Government Board had that year suggested that Medical Officers of Health should make a thorough inspection of the encampments before the season began and report any infringements of by-laws. He reported some of the comments of the Medical Officers of Health of the county districts. In Tonbridge Rural District, 199 farms were visited before the season started and a number of alterations were made before the hoppers arrived. Bell-tents provided by some farmers were strongly condemned, and 150 cubic feet of room-space was recommended as the minimum for each adult.

The problem of providing suitable toilets for the two sexes was recognized, but in the Tonbridge area farmers were recommended to make up simple moveable structures in pairs, with a framework of laths covered with sheet iron, and so designed to be moveable on poles sideways over a trench about 2 feet 6 inches deep by 1 foot 6 inches wide. The idea was that dirt from the trench could be thrown back in as necessary, and when that part of the trench was thus filled level with dirt again, the whole structure could be moved sideways along the length of the trench and the process begun again.

A Local Government Board Inspector who toured various hop-growing districts of England between 1911 and 1913 made a number of specific recommendations, most of which had been made before. He, too, condemned tents as "not suitable for occupation by women and children". When used at all, he said, it should be by men and boys, and the tents should be "of sound repair and thoroughly watertight". In specially built huts, all the windows were to open, there were to be proper guttering, paths, a few shelves and pegs. Beds should be raised above the ground, unless the floor was of impervious material, and where wooden frames of beds were not provided, then the bedding should be raised on hurdles or hop-bines.

What may have been a new feature of all the many recommendations that were made over the years was that in each camp someone should be appointed by the grower to look after the sanitary conveniences, and a woman should be appointed, too, to take charge of young children and infants who were left behind in the camp while the parents were in the hop garden, picking hops.

The First World War hustled in a great many changes to life in Britain generally. It brought a shortage of hop-pickers to Kent farms. Women who had previously poured into Kent to find casual labour to help eke out the family income, now found better conditions and better pay nearer home, doing war work of one kind or another. Many of the men were in the Forces, and as the war went on and conscription was introduced, this became more and more marked.

The hops still had to be picked, though. The war did very little to the popularity of beer—unless it increased it! In 1917 and 1918 hopping suddenly acquired a quite new respectability. From being work for the lowest levels of society only, it became another job of war work and there was an invasion of volunteers into the hop gardens, organized by schools and colleges. For the first time, hopping was socially acceptable—if only for the duration—and the sense of holiday among the hops crept into the annual event on a wider scale than ever before.

In 1917, the hop acreage was down by more than 50 per cent anyway, so the inexperienced young volunteers were able to cope quite well. The reduction in acreage was ordered by the Board of Agriculture and a Controller of Hops was appointed in 1917 to regulate the quantity of hops coming on to the market.

When the war was over, many employers expected life to revert to the way it had been before. In the 1920s, with the hop acreages increasing again, growers preparing for the old-style invasion of hoppers from London still emptied wagon and cattle sheds for the visitors. Standards of cleanliness were much improved from the old days, and it was usual now to lime-wash the walls of the sheds, but the floors were still considered to be adequately furnished with a deep layer of straw and old hop bines. Sacking partitions were put up to separate one family from another, and toilets were provided—usually two wooden or corrugated iron 'sentry boxes', each with a wooden box with a hole in it set over a deep hole in the ground. On the door of one would be painted the word MEN; on the other, the word WOMEN.

A communal cookhouse took the form of some iron bars placed across loose bricks on the packed-earth floor beneath a tin roof. There were no walls to the cookhouse. Faggots of wood for the fires were provided, though not always very conveniently near the encampment because farmers knew that the shorter the distance separating faggots and fireplaces, the faster the wood was burned.

Certainly, things were better in the 1920s, though not sufficiently better to escape the criticism of the County Medical Officer of Health in his Quarterly Reports all through the decade. The criticism told of poor sanitary facilities, danger from open fires, the risk of the spread of infection, and the need for better medical attention in the camps.

In 1924 Mr S. Nicol Galbraith, Medical Officer of Health to the South West Kent United Health Area, reported:

> "I have seen the slums of Dublin and Salonika, but conditions here put these in the shade. A picker stated that the latrines of these permanent brick huts had not been emptied for three seasons. Little wonder epidemics broke out and all credit is due to the voluntary workers, many of them trained nurses with war service, who have worked in such conditions, not for gain or even glory, but for humanity."

Yet, a year later, a party of Ministry of Health doctors visited Kent for twelve days, at the end of which they reported:

> "It is evident that the lot of the Kentish hop-pickers during the past two years have been greatly improved. Public opinion is helping local health authorities in their efforts to raise the general standard of camp conditions."

Galbraith was still referring to "gross overcrowding" in 1925—and not without cause, it seems. In one hut he inspected, measuring 12 feet by 10 feet, he found no partition separating four families of nineteen people. In another, smaller hut, he found three families of fifteen people altogether. Things had not, apparently, changed all that much, at least in some places, in a century.

"With my Inspectors," Mr Galbraith reported, "I visited these huts after 9 p.m. The crowded, unhealthy conditions could scarcely be described with words in the English dictionary."

At weekends, things were inclined to be worse. It had been a long tradition of the hop gardens for hoppers to have visitors—relations and friends from London—down for the weekends and if they stayed overnight (as they did) then it was a case of: "Move over, family; room for two (or three or more!) more."

But, whatever the conditions they found when they came here, the hoppers still poured into Kent every year, and treated the whole adventure as a working holiday. By now, life in the hopper huts was undeniably rougher than that which most of them lived the rest of the year in London. But if the accommodation was crude and the facilities minimal, the Kentish hop gardens still offered, as they always had done, three weeks of aromatic fresh air during what is often the best season of the British year, with a chance to earn good money in one of farming's most pleasant harvestings.

Miles Sargent, in his book *St Francis of the Hop-Fields*, gives a vivid description of what he found when he visited a hoppers' hut one night in 1926.

"A dim oil lamp hung on a nail on the wall," he wrote;

> "rows of children, some arranged this way and some that, on the huge mattress, some asleep, some half awake, all almost completely dressed, and the all-pervading stench of dirty humanity, pickles, hops, dirty clothes, babies and paraffin."

In his book, Miles Sargent called the hoppers "the most amazing mixture of courage and fear, of comedy and pathos, of patience and hysteria, indeed of almost all the contradictories in human nature".

They came, he said, from the very lowest parts of London, and every slum area supplied its quota. Few of the men were in regular work (remember, this was 1926); many worked in the docks and did not know from day to day if there would be work for them or not. But although they came from some of the toughest and roughest parts of London, most of them had experienced exploitation all their lives and were resigned to it. They were, almost all of them, under-nourished (though not necessarily under-fed) and they had learned to expect to be swindled and cheated by authority in all its guises, not really because authority set out to swindle and cheat them, but because authority devoted very little of its time to understanding them.

Because authority treated them badly, they in turn treated each other badly: their tradespeople took advantage of the customers' simplicity and lack of learning, and the customers took advantage, whenever they could, of the tradespeople's inability to watch all the goods all the time.

They were largely untaught, unsophisticated, generally unloved—at a time when to be any of these things was still regarded by society generally as the fault of the individual for not trying harder. They were polite society's left-overs in their own city; why would they expect to be treated as anything different on the farms and in the villages of Kent where they did not belong?

They came hopping for the chance it offered of a bit of fresh air and the nearest many of them ever came to having a holiday, and to be paid for it, but not all of them were happy about it. Many of them didn't really like the country at all, with its unlit, high-hedged lanes, and its unaccustomed night-time quiet. They moved about, from the huts to the nearest pub and back again, in the evenings in packs, singing at the tops of their voices as much to bolster their courage as because they were drunk. They frequently got lost. But they were, above all else, survivors. They were realists, with the instinct to look after Number One in whatever circumstances Number One found himself at any time. They were not good at making the best of things, but they were very good indeed at accepting the worst of things.

In Father Sargent's own words: "They are so easily led that but little honour attaches to those who spur them on to deeds of violence, for they are so pathetically dependent on any one who seems to them to have even slightly higher intelligence than they have."

He noted the Londoners' tendency to stand and watch rather than act in cases of emergency; he judged this to be because, in London, no-one ever asked them for help, and, in fact, authority always tended to hold them back behind a cordon whenever any kind of emergency arose. He was impressed by the natural decency of the pickers, in spite of the conditions in London, so that although they were direct and called a spade a spade, they were not vulgar or immodest.

Father Sargent obviously was one of those hop garden welfare workers who inspired respect and restraint among 'his' hoppers, and that no doubt accounted for their manner in his company. Nevertheless, anyone who ever shared a hop garden with a crowd of London hoppers, must surely wonder if he did not, perhaps, close his eyes and ears against some evidence that his judgement of them was not, to say the least, open to contradiction.

The old relationships between the visitors and the native villagers changed very slowly indeed. Village shops still took anti-pilfering precautions before the hopping started. They used to put wire netting over their displays, and they moved the more obvious temptations out of easy reach. Local pubs, too, brought out reserve glasses which were specially thick, like cut-down jam jars, for the hoppers each year and sometimes the beer on offer to the hoppers was of a quality that the publicans would have hesitated to offer to their regulars. They could afford to let standards fall a good deal; during hopping, beer was definitely in a seller's market. For the Londoners, the consumption of a great deal of beer was an essential part of hop-picking. Their measure of a good hopping was often the quantity of beer they drank during the season.

Hop-picking time was always a period of specially good business for all the hop-growing areas of Kent. The hoppers had to be fed as well as supplied with beer, and when hopping was over, many of the hoppers spent a good deal of the money they had earned in the larger towns—Tonbridge, Maidstone, Ashford.

By the 1920s, hopping was firmly established among many families, foreigners and home-dwellers, too, as a time for re-equipping the family, especially the children, with winter clothing, and if there was any money left over after that, then the parents might come in for a share, as well. It was by no means uncommon for three or four cartloads of discarded old clothing to be cleared away from streets and doorways in Maidstone at the end of a Saturday's shopping. People came in from the hop gardens, bought themselves new clothes, and simply changed into them on the spot, throwing the old ones away as they did so. The cost to the borough for collecting and disposing of the rubbish was part of the price paid for the extra business done by the local traders, and the town regarded itself, on balance, as well served by the annual visitors to the surrounding rural areas.

In 1926, the Minister of Health issued new model by-laws controlling hoppers' conditions, bringing them more into line with current standards, which had improved a great deal in a quite short space of time. It was not always easy for farmers to comply with these new standards even if they felt inclined to do so. Vandalism was a common problem and the old Cockney aversion to 'draughts' made it almost impossible to ensure that huts were properly and healthily ventilated.

There were now generally enough hoppers' huts, but there was still a good deal of overcrowding in them. The by-laws laid down recommended sizes for the huts and for the rooms or apartments in them but, as almost always happens when recommendations are enshrined in law, these were generally taken to be 'legal' sizes rather than suggested minimum ones. So all the huts tended to be of the same size, whereas, of course, all the families of the hoppers were not. And since families usually wanted to stay together, it was common for some of the huts to be quite seriously overcrowded, especially at weekends.

Some attempts were made to stop the practice of weekend visiting, but they were resisted very robustly indeed by the hoppers themselves, and the growers feared there would be a return to the old days of rioting if any such restrictions were imposed.

One of the continuing contentions of the Kent authorities was that the London hoppers brought to Kent expensive problems for the county's public services, especially the health services, and there were many arguments that called upon the London County Council to contribute towards these costs. It was pointed out that after a stay in the Kent hop gardens, London school children returned to the capital with much improved health, and that this was recognized in London itself, where schools were closed specially so that the children could go hopping with their families.

Kent's argument with London ran along these lines: "Since most of these children are too young to be of much help to Kent hop growers as hop-pickers, the only point in their coming is for the good of their health. Kent is, therefore, doing a service to thousands of Londoners by providing what are virtually cheap holiday camps for them."

The argument went on to expect that London would recognize the logic of contributing towards the cost of that holiday for its young dependents. As the county said on one occasion: "We in Kent do not wish to discourage the Londoners from coming, but we aim at the migration being properly controlled in the mutual interest of the public health of both Kent and London. These poor people will therefore continue to enjoy their only holiday in our beautiful county under the best possible conditions."

London, not altogether unnaturally, tended to take the view

that these 'poor people' would not have gone hop-picking at all if the Kent farmers and hop-growers had not needed them so badly and that, therefore, it was they who were the benefactors, rather than vice versa.

In March 1931, Dr Galbraith and Dr Constant Ponder visited the offices of the London County Council to discuss the question of control of infectious diseases. Sometimes, they pointed out, infected hoppers could not be isolated from the rest; rather than be left behind in isolation in Kent, they escaped back to London by train or bus, risking the spread of infection. Dr Galbraith said he knew of fifteen cases of smallpox that had travelled back to London from Kent in this way.

In her report, Dr Ponder said:

> "The conference was unsatisfactory and produced no offer of co-operation. The Kent Health Authorities have done all in their power to effect agreement with the LCC but without result. Should, therefore, any untoward event or epidemic arise in Kent or London from the uncontrolled shuffling of cases of infectious disease among hop or fruit pickers between London and Kent, the London County Council will be entirely responsible."

The argument was never properly resolved. Perhaps there was no way of resolving it. The numbers of people involved were very large, and the people themselves did not take at all kindly to any sort of interference with or attempt to control what they regarded as very much their own affairs.

Running through all the reports of everyone in authority who commented on the situation were constant grateful references to the voluntary organizations without whom the health of the hoppers might well have been very much worse than it was.

Voluntary workers were a regular feature of hopping for many years—certainly from the 1850s. They tended to offer somewhat specialized services to the hoppers, but to some extent they all became involved in both welfare and medical work at some level. During the hungry thirties, it was the volunteers who provided field canteens, serving tea and buns at rock bottom prices. They organized crèches for the babies and first-aid posts for all and were always on hand with advice and help of all kinds.

This, of course, was before the days of the welfare state, and although the social conscience was far keener now than it had been ever before, there were still great gaps through which whole sections of the public could and did slip. Annual contact with the hop-garden volunteer welfare workers was sometimes quite literally a life-saver for people who might otherwise never have made such contact from their own homes.

By 1931 the acreage of hops in South-West Kent alone was 6,706 acres on 386 farms. Between them, these farms had 11,548 purpose-built huts which, together with some temporary tents still surviving here and there, accommodated about 60,000 hoppers. The total for the whole of Kent would have been, perhaps, half as many again. The annual report for 1931 of County Medical Officer Dr Arthur Greenwood noted that water was laid on to 296 of those 386 farms, and he commented: "This is a better percentage than is the case with dwelling houses in many rural districts."

That annual report, like many others, praised the amount of voluntary work being done for the pickers. Dr Greenwood called it "remarkable".

"In the South-West Kent area there are two hospitals, 47 dispensaries, serving 309 camps," he reported, and he went on to say that the volunteers treated 6,622 new patients that year who, during the course of the season, made 15,268 attendances at the voluntary centres. Of them, 154 were transferred to hospital for more care or treatment than the volunteers could offer. The fact that the bodies were 'voluntary' does not mean they were 'amateur'. They employed four full-time doctors between them, twenty-five trained nurses and seventy-three part-trained workers. The Area Medical Officer added his opinion that the pickers in Kent were, medically, "very well looked after".

A lot of the work of the volunteer medical workers was comparatively minor. The doctors and nurses were kept busiest with patients suffering mainly from simple unfamiliarity with the ways of the countryside. There were always stings by wasps, bees and other insects, usually more painful than serious, but always potentially more serious for being neglected. Children who had never climbed trees before often fell out of them; children who had never seen a stretch of unfenced water bigger than a storm-water puddle in the street were specially likely to fall into ponds or even creosote vats. Burns and scalds were very common; infected or inflamed eyes—known as 'hop eye'—were also common; and so were hop rashes, allergies, stomach upsets, and so on.

There were more serious cases, too, of course. The visitors did not always know the difference between edible and poisonous hedgerow fruits; occasionally there were accidents resulting in broken limbs or heads. Cuts and sprains were not always of a minor character. Life in a hop garden dispensary could be very varied indeed, and a very good training ground for student doctors and nurses.

The Labour Government that was in power in 1931 appointed inspectors to see that all hoppers had proper accommodation and if there was none, then the farmer was only permitted to employ local pickers. It was, no doubt, a well-intentioned move, but it had the effect of barring hundreds of unemployed people, who might otherwise have found work in the hop gardens, from doing so, and it was bitterly resented by the hoppers as much as by the farmers.

At this time, Britain was in the lowest trough of the great pre-war depression. In the true tradition of 'it never rains but it pours' not only were many unemployed people unable to come hopping in Kent because the available accommodation did not come up to Government required standards, but also the hops were poor, the picking season was wet and hop prices were very poor, too.

As a result of the boom in hop production during the previous century, and because there was an influx of cheap imported hops, too, there was a sharp and continuing decline in acreages of hop gardens to about 17,000 acres by 1932, by which time hop growing in England was in a truly perilous state.

The first Government control of hop production had been introduced in 1914, primarily to help the industry during a particularly difficult period. Then, beer duty was raised from seven and ninepence to twenty-three shillings a barrel. The result of this was a hop surplus and growers petitioned the Government to put a duty on imported hops, which it did in 1916.

In 1917 the Government restricted brewing to just over a quarter of the pre-war output and this led to a massive hop surplus. Growers urged the Government to order reduced acreages of hops, and in fact the acreage went down to about

Hoppers waiting for a special train at London Bridge station in 1935

The horse-drawn wagons carrying the hoppers from the station to the farm will later be used to transport the hops to the oast houses

There was always a good deal of free-and-easy informality at the bin-side, as this 1937 picture shows

A British Red Cross Society dispensary at work at East Farleigh

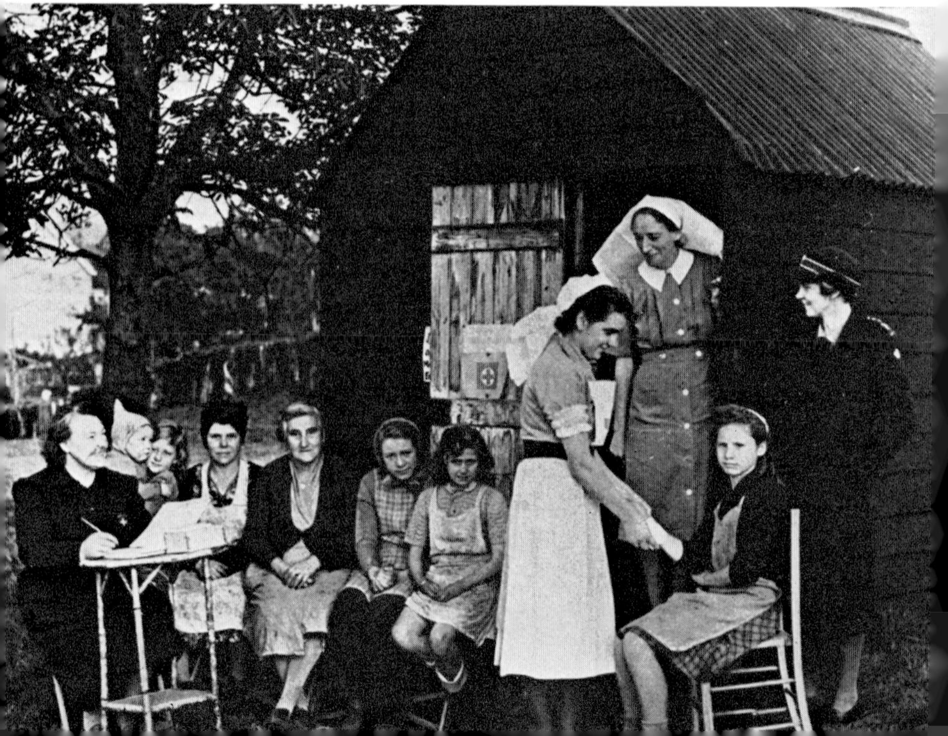

half the 1914 crop. It was in 1917 that the Hop Control was introduced, with a committee appointed by the Board of Agriculture under the chairmanship of Mr G. Foster-Clark.

But even then large stocks of hops accumulated. The situation was made worse by the fact that the 1924 crop was the best ever, producing 477,000 hundredweights of hops, and in 1925 the Hop Control ended. A duty of four pounds a hundredweight was slapped on imported hops and the English Hop Growers Ltd, a co-operative formed by most of the 1924 growers, was set up. It was not successful. Non-members sold their hops while the growers were still deciding what value should be put upon them, and in following years, members fell out of the co-operative which was finally wound up in 1928.

The Hops Marketing Scheme was introduced in 1932. Basically this was a scheme that restricted growers to selling all their hops only to the Hops Marketing Board, which was administered, predominantly, by growers themselves. Two years later the quota system was introduced, under which an annual quota was decided by agreement between the Board and the brewers, and announced to the growers, who thus knew how many hops were going to be needed in any year and could plan accordingly.

The Hops Marketing Board was first established in London, but is now based at Paddock Wood, in the very heart of hop-growing South-West Kent. It represents the English hop industry at all levels within the European Economic Community and carries out an important and, indeed, essential function in this way. It also plays a leading role in the affairs of the International Hop Growers' Convention, which represents all hop-producing member states, and is particularly concerned with the problem of balancing international supply and demand, and helping growers to combat the effects of inflation.

The Board brought stability to the industry, and following its creation acreages of hops rose to about 20,000, a figure that held steady for many years. More recently, though, changes in brewing processes and changing tastes of the drinking public brought a reduction to about 17,000 acres.

It has been claimed that without the introduction of the Hops Marketing Scheme, hop-growing might have ceased in this country altogether, and brewers might now be depending upon imports from overseas. There are changes ahead, however, and the Hops Marketing Board is even now taking a hard look at the future to see how it can adapt its production and marketing techniques to meet the new conditions.

The scheme was still a year away, though, when, in 1931, George Orwell went hopping down in Kent and the essay that he wrote afterwards about the experience has left us with a very vivid picture of what hopping was like then.

Orwell quite enjoyed the adventure, although he admitted he felt a wreck at the end of it, and he felt humiliated that most of the other people in the hop garden looked upon it as a holiday.

Orwell came to Kent with a twenty-six-year-old companion called Ginger, an ex-Borstal boy and burglar, a twenty-year-old orphan they all called Young Ginger, and an eighteen-year-old Liverpool Jew. They were all living rough in London when they decided to make their way together into Kent for the hopping.

The writer's social concern shows through much of his account of the expedition. He observed that laws about child labour were disregarded utterly, and some people drove their children pretty hard in his opinion. He also commented on the habit of 'heavy' measuring that was practised by some measurers. Remarking that there are twenty bushels to a full bin, he added that if the hops were measured 'heavy' a full bin could be booked at only twelve or fourteen bushels.

Theoretically, he reckoned, a picker could earn about thirty shillings a week, but he claimed that no-one actually did earn that much. He personally doubted if any one picker made more than fifteen shillings a week, and he believed that it was because hopping was regarded as a holiday that the pickers would take such 'starvation' wages.

"It gives one an insight into the lives of farm labourers, too, to realize that, according to their standards, hop-picking is hardly work at all," he commented.

The farm where Orwell picked was paying a tally that year of six bushels a shilling, but that was quite good. Nearby farms were paying only a shilling for as many as eight or nine bushels.

In 1931, pickers on arrival at the farm were handed a printed copy of the rules of the garden, which Orwell described as "designed to reduce a picker more or less to a slave". The rules gave the farmer power to sack a picker without notice, and pay him off at eight bushels a shilling instead of six—in other words to confiscate part of his earnings. If a picker left the job before the picking was finished, his or her earnings were docked by an equivalent amount, in fact, about a quarter.

Pickers were not allowed to draw what money they had earned up to any particular time and depart because the farmer would not pay more than two-thirds of any one picker's earnings in advance, so that pickers were always one-third in arrears at least until the final paying-off. Binmen were paid weekly wages rather than piece-work rates like the pickers, but if the pickers came out on strike, then the binmen's wages ceased until the strike was over, so it was in the interest of the binmen to do whatever they could to help settle hop garden disputes before they reached the stage of a stoppage of work.

Orwell advocated a pickers' union to give them strength against the employers, but he recognized there was little point in trying to form such a union while about half the pickers were women and gipsies "too stupid to see the advantages of it".

The farm where Orwell worked accommodated the pickers in round tin huts about ten feet across. There were windows, but they had no glass in them, and they were not weather-proof. For furniture, they had a heap of straw and hop-bines (Orwell himself called them hop-vines) and nothing else. He shared his hut with three other people, but he reported that some of the huts had seven or eight people in them. He thought that sort of crowding was something of an advantage, because it at least kept them warm.

There was free firewood for the pickers, though not, in his view, enough, and the water tap was 200 yards from the hut. The nearest latrine was also about 200 yards away, although Orwell described it as so filthy that he would have walked a mile sooner than use it. He said the pickers used a nearby stream in which to do their laundering but, he added, getting a bath in the village would have been about as easy as buying a tame whale.

Orwell's hoppers were up at six o'clock in the morning to be ready for a six-thirty breakfast of tea and fried bread. They made sandwiches and bottled cold tea to take with them for their dinner at the bin-side. They picked hops all the morning until 1pm, when they lit fires between the bins to heat up the tea and ate their sandwiches during the half-hour meal-break. Then they went back to work, picking until about 5.30pm.

The Saturday night tradition of high times around the village pubs was little affected by the depression. Drunkenness was as common as ever; police often had to be called in to clear the pubs at closing time and sort out the fights that broke out. Orwell commented that he supposed the villagers must have thought the pickers a nasty vulgar lot, but he wondered if it was not also good for a dull village to have this annual invasion of Cockneys.

Orwell's experience of hopping and the account he has left of it is interesting and valuable. But it has to be said that he could hardly have chosen a worse time during all the twentieth century to go hopping. The depression was at its deepest, the acreage of Kent hops was under 9,000 acres (still more than half the total acreage of hops in England and Wales, incidentally) and had been dropping for the past twenty years. One of the difficulties of accurately assessing hopping conditions at any time during the last 100 years at any rate is that many of the people who have left the most graphic accounts were people with a lively social conscience, not always in step with general public opinion. Many of them were actively involved in bringing about improvements to conditions in the hop gardens and they were, therefore, anxious to recruit support and help. There is no doubt at all that some of them, while not actually misrepresenting the true position as they saw it, did make rather more of the worst conditions they encountered than they did of the best they found or heard about.

Orwell's account, for instance, is undoubtedly generally accurate. Yet we have accounts of ordinary hoppers of the same period recalling that they were at their bins, not at six o'clock in the morning, as he said, but at seven. And certainly not all the hoppers, even in 1931, went to work on a breakfast of fried bread and tea. Some, recalling these days later on, remembered with something close to wistfulness, substantial breakfasts, followed by mid-morning wedges of cake and bread pudding. Simple fare, yes, but far from inadequate.

Although he reckoned that no one picker earned much more than fifteen shillings a week, other accounts of hop-picking in the 1920s and 1930s tell of two people at a bin being able to earn up to ten pounds for the whole hopping. In this matter, a great deal depended upon the expertise and experience of the pickers, and also, perhaps, upon how accurately the memories of some of the other hoppers recall how much casual help they had at the bin from day or even half-day visitors who, of course, came to the hop garden for the experience, picked many or few hops according to their inclinations and abilities, and left the bin-holders to add their contribution, unrecorded, to the end of the day's measure, and the end of the hopping pay-off.

It is a fact that people who are recalling how pleasant it all was do tend to forget such things; people who were intent upon emphasizing the need for improvements do not.

Later in the 1930s, of course, things generally got better. In 1933, Dr Galbraith advocated a levy of (he suggested) one penny on each hundredweight of hops picked for the growers to provide improved accommodation and medical care for the hoppers.

"It does not seem right," he pointed out, "that such a wealthy industry as hop-growing should allow itself to be dependent upon charitable organizations for the welfare, medical and nursing work done for its employees, the hop-pickers."

And a year later, in 1934, County Medical Officer Arthur Greenwood was commenting: "I feel now the financial position of growers has been put on a sound basis and it is certainly no longer possible to plead poverty as a reason for not carrying out essential improvements, that local authorities should deal sternly with delinquents." District councils, he said, were still not enforcing their own by-laws.

At the same time, though, both he and Dr Galbraith were forced to confess that not all the blame could be laid upon either the local authorities or the farmers. Although Dr Greenwood was lamenting the serious shortcomings in refuse and litter disposal in the hopper camps, and saying that many growers still did not understand that unless reasonably decent clean and comfortable provision was made for latrines, women and children would not use them, and illness would result, Dr Galbraith had to point out that even though refuse bins were now, at last, being provided in almost all the camps, many of them were being misused by the hoppers.

"Last year," he reported in 1934, "we experienced the same old difficulties of getting pickers to take even a minimum of trouble to look after themselves. We found refuse bins provided by the growers being used as anything but for what they are intended."

Similar comments applied to the cookhouses which were required, by local by-laws, to be provided at the camps. Dr Galbraith said that at one farm solid brick fireplaces had been pulled to pieces so that the bricks could be taken away and used to make open fires all round the camp. The hoppers preferred it that way, despite the obvious risk of fire and the scores of children who were burned or scalded every hopping, sometimes with fatal results. And, of course, any housewife will have some sympathy with this preference—certainly, any housewife who has ever had to share a kitchen with even one other family will understand it.

In 1934, though, Dr Galbraith was able to report that some of the camps in the Maidstone rural district were so improved over the previous year that "we could hardly recognize them". In 1937, he said more than 500 new huts had been built in South-West Kent, and many more had been improved. New and extended cookhouses were at last being used as they were meant to be used.

In the last of the County Medical Officer's pre-war reports, in 1938—and, incidentally, the last ever to make special reference to hop-pickers' camps at all—Dr Constant Ponder was able to say that infectious diseases notified had dropped to eighteen (compared with last year's twenty-four) and that none of these had been in the Maidstone rural district—probably the busiest hop-picking district in Britain.

SEVEN

The War Years—1939–45

There were hops, hops, big as turnip tops
In the store, in the store;
There were hops, hops, big as turnip tops
In the Quartermaster's Store.

My eyes are dim, I cannot see,
I have not brought my specs with me—
I have not brought my specs with me!

Hoppers' version of a popular multi-verse song during the Second World War

The thousands of hop-pickers who streamed out of London into Kent in September 1939, as usual, did so with an even greater-than-usual sense of relief and relaxation. War and talk of war had hung heavily over the capital for months, years even, by then. But in September the waiting and the uncertainty were over. By the time hopping began, Britain and Germany were at war, and London seemed a very good place to be out of, if only for a little while.

The special trains brought the hoppers into Kent in almost their customary numbers, although there were rather fewer men than usual this year because many of them had been called up and were already in uniform. Even so, a lot of them were able to get weekend leave and help their families move into Kent before returning to their barracks and shore stations.

Some of the serving men were even able to get special leave to go hop-picking themselves (this, remember, was the 'phoney war' period) and others found themselves detailed into the hop gardens as part of their fatigue duties.

Everyone who did come had, for the first time, one new item of luggage—a gas-mask; and many of the huts they came to were now camouflaged against the possibility of air attack.

In Kent, they found a good crop of hops, a tally of about three bushels a shilling, and, one way or another, it promised to be a good hopping, war or no war.

Although the reminders of war were everywhere, even in the hop gardens themselves—the gas-masks, the camouflage, and the uniforms—the realities of war were yet to come to England. There were all the usual features of the annual hopping down in Kent: the mobile traders still toured the gardens, and the voluntary workers were still there, although fewer in number.

Many of the volunteers of previous years were also turning their energies towards helping the general war effort, and the first-aid posts were short of trained nurses, many of whom had gone to work in the hospitals or the military nursing services.

By and large, however, the first hop-picking of the Second World War was very little different from the last one of the between-wars peace. There were many pickers, of course, who remembered picking all through the First World War when, apart from the occasional Zeppelin sighting, and the inevitable food and other shortages, the war had almost wholly by-passed the Kent hop gardens, at least as far as the pickers were concerned.

In 1939 there seemed to be no real reason to suppose that the same thing would not happen this time, too. After all, as many people confidently predicted, it wouldn't last very long. The Germans might have blitzed Czechoslovakia and Poland, and got away with it, but now they had Great Britain and her Empire to contend with. We'd licked 'em before; we'd soon show them we could do it again.

And so, despite a labour shortage that, just as it had only twenty-two years earlier in the closing half of the First World War, brought schoolboy and student volunteers into the hop gardens to help harvest the crop, hopping went on very much as it always had, war or no war.

The possibility of air attack was fully recognized, though, and many of the hop gardens were flanked with specially dug trenches, some of them lined with sandbags and roofed over with sheets of galvanized iron, to provide air-raid shelters for the pickers if they should be needed.

Air Raid Precaution Wardens were kept busy checking the blackouts at the hopper huts. Not only did the huts themselves have to be carefully blacked out, but cooking fires had to be doused before nightfall, too and, of course, the oast houses had to have sacks and blankets over their doors to make sure no hint of the fires within seeped out into the night. Apart from these really quite minor inconveniences, the 1939 hopping season ended without any of the incidents that many of the hoppers and growers had prepared for.

The 1940 season, however, was very different.

After June 20, 1940, no-one was allowed into certain designated restricted areas of Kent unless they legitimately belonged there. The areas covered a large part of the county, including the City of Canterbury, the boroughs of Chatham, Dartford, Deal, Dover, Faversham, Folkestone, Gillingham, Gravesend, Hythe, Lydd, Maidstone, Margate, New Romney, Queenborough, Ramsgate, Rochester, Sandwich and Tenterden.

Also restricted were the urban districts of Ashford, Broadstairs, Herne Bay, Northfleet, Sheerness, Sittingbourne, Swanscombe and Whitstable; and the rural districts of Bridge-Blean, Cranbrook, Dartford, Dover, East and West Ashford, Eastry, Elham, Hollingbourne, Maidstone, Malling, Romney Marsh, Sheppey, Strood, Swale and Tenterden.

In other words, all of north and east Kent and much of Mid-Kent as well. Many farmers, especially in these restricted areas, feared a severe labour shortage for the hop-picking. In East Kent, where the dependence upon Londoners was less extreme, these fears were fairly minimal, but the Mid-Kent, West Kent and Weald farmers needed at least 50,000 Londoners a year to supplement the home pickers.

The Kent farmers' fears were voiced in Parliament more than once as the hop-picking season approached. But by August the crisis had sorted itself out. It was agreed that, subject to the military situation at any time, the Londoners would be allowed into the restricted areas for the picking. Children, too, were to be allowed to accompany their families into the West Kent areas, but not into East Kent. The hop crop was saved.

All through that long, clear-skied summer of 1940, the Battle of Britain had raged, day and night, over the southern counties battleground. By September's hopping, the battle was in full swing, and when the Londoners came to Kent they left one front line for another.

The Londoners generally felt that the Kentish hop gardens

were a whole lot safer than the streets of London, from which some of them had already been bombed out, some more than once. They were issued with emergency ration cards to bring with them, and many brought hoarded tinned foods—just in case—and they made their annual pilgrimage to the hop gardens partly because of, partly in defiance of, the war.

The air-raid-shelter trenches were certainly needed this year, and picking was organized so that hoppers were never too far away from cover of some kind. The pickers themselves organized their own air-raid precautions, and some of the older children reinforced official 'spotters' whose job it was to give the alarm whenever the overhead battle came specially dangerously near.

Air-raid-warning sirens sounded daily—often several times during the day—and every night, too, was punctuated with the explosions of bombs and gunfire. During the day, hopping was often interrupted by dog-fights between the RAF Spitfires and Hurricanes and the German bombers and their fighter escorts.

The hoppers shared with the natives of Kent a grandstand view of the Battle. The late summer skies over Kent were a sparklingly-clear blue, almost as though they had been specially valeted for the spectacle.

Day after day, the dome was scarred with the trailing vapour lines, like claw marks in flesh—close-set and regular—behind the tiny silver crosses of the invading bombers.

"There they are!" the cry would go up, and sure enough, out of the sun an even smaller pin-head of brilliance would trail its own individual scar down into the pack of bombers. Others would follow, weaving and plunging, breaking up the tight formations and tangling the neat vapour patterns in the silent slow-motion of mile-high combat. Here and there, planes plummeted, exchanging the lazy white streamers of the packs for solitary plumes of black smoke or great balls of fire out of which fragments fell in all directions.

When the engagement was overhead, as it often was, the hoppers went to the trenches, and when it had passed for the moment, the children went in search of trophies: cannon shells, cases and twisted fragments of the machines of friends and foes alike.

Sometimes the victim of a dog-fight would drop to hedge-hopping height and streak for home, emptying his guns into the fields and villages as he went. Then there would be no time to run for cover, and the hoppers dived under their bins and hoped or prayed as their natures dictated. The bins, even full of hops, gave them no real protection at all from the cannon shells and machine-gun bullets, but at least they hid the bright colours of the aprons and dresses of the women from the bandit marksmen, and may have saved some of the gardens from a deliberate straffing.

Everyone who went hopping in 1940 ended the season with his or her own tale to tell. For some it was of bombs that just missed their hop garden and fell in a nearby field. For some, it was of crashing planes skimming the bines so low that they could clearly see the crews leaping from the fuselage doors and feel the heat of the flames that flared from the silent engines.

Enemy fliers parachuted into fields and woods and were taken prisoner by pole-pullers with their wicked-looking hop-dogs until they could be handed over to Home Guards or soldiers. RAF pilots who came down in the same way were assured of a quite different welcome.

Sometimes, hoppers were machine-gunned as they fled from their bins for the nearest cover—a trench or a ditch, or a nearby railway bridge. At least one group, bolting along a cart-track for a branch-line railway bridge, reached the cover only feet ahead of the bullets thudding into the track behind them, and if the distance had been only yards more than it was, and the pursuer could have kept his course seconds longer, some of them would certainly not have lived to remember it.

Not all the hoppers that came to Kent that year were so lucky, inevitably, and there were casualties, although not as many as there might well have been.

Not for centuries had so many civilians had such a grand-stand view of a major battle being fought—certainly not on British soil, and even more certainly not over it.

When each dog-fight was over, and the All-Clear sirens wailed their sighs of relief again, there was usually one Hurricane or Spitfire to over-fly the area and to do a low-level victory roll or, at least, to waggle its wings in acknowledgement of the congratulatory waves of the hoppers below, before they went back to snatching the hops off the bines and cursing Hitler with whole-hearted Cockney venom for daring to interfere with their hopping.

The labour shortage of 1939 was much worse in 1940, and far more local people were recruited into the hop gardens. People who, before the war, had always regarded hopping as socially quite unacceptable work were persuaded to overcome their scruples in the national interest and give it a try. They found that hopping was a far more specialized job than it looked or sounded and that it took years of regular hopping to become as proficient as some of the older pickers undoubtedly were. Still, every little helped and with the added help of soldiers and the girls of the new Land Army, the hops were picked.

In some villages, so that hoppers would not have to run the risks involved in working out in the open hop gardens, most of the time unavoidably some distance from even the most minimal protection from the overhead battles, the hops were brought to the village streets to be picked there. Bins were set up in the street, and the men went out and cut the bines and brought them back by the wagon-load to be picked.

It was not a good way of harvesting the hops. Healthy hops as they hang on the growing bine are crisp and firm to the touch, and can be snatched off fairly easily by experienced fingers. But within a very short time of the bine being cut, the hops soften, and begin to wilt. Then, instead of snapping off the bines under the deft wrist movement of the pickers, they tend to break and powder much more easily, so that the stripping action that can clear a branch of hops in one movement, has to be abandoned and each hop has to be plucked separately.

More strength is needed for each hop, too, so it is both more tiring and slower to pick hops from bines that have been pulled a little while, especially if they have lain in the hot September sun during that time. But bringing the hops to the villages did mean that local pickers could avoid the trek to the gardens night and morning, and if there were aerial activity to interrupt the picking for a while, the villagers could scatter into the comparative safety of their own homes, until it was safe to come out again and carry on picking.

With the tally at about three bushels a shilling in 1940, many hop gardens had their own Spitfire Funds—or rather their own schemes for contributing to the Kent Spitfire Fund which existed to raise money to buy Spitfires. A typical scheme meant that the growers paid the pickers a shilling for every two or three bushels (according to the local tally) and gave another sixpence a poke (of ten bushels) for the fund. There were collections among the pickers, too, for extra contributions to the fund.

The greatest danger during the Battle of Britain to Kent villages was not from bombs (most of which were, anyway, intended for London area targets rather than the Kent towns and countryside) but from anti-aircraft shrapnel, falling shell-cases, crashing and disintegrating aircraft but Kent was bombed as well. Newspaper reports at the time were obliged to be guarded about where incidents occurred, but at least one lot of hopper huts were bombed and between forty and fifty were completely shattered in October 1940. Hopping was actually over for the year when that happened, but some of the hoppers were staying on in the huts for the weekend before returning to London.

Another feature of the 1940 hopping was that although the regulations applying to restricted areas were relaxed for the hoppers themselves, special permission was needed to be able to visit them and so the tradition of hoppers entertaining many times their own number of visitors during weekends was interrupted. Only one visitor from each household was given a visitor's permit at any one time.

The winter of 1940–41 was a specially harrowing one for Londoners. The Battle of Britain was over and won, but the Blitz on the cities—first London, later the provincial cities—raged on. Newspaper and radio reports and pictures left the rural dwellers profoundly thankful for their comparative safety, and marvelling that the Londoners were able to stand up to such a hammering.

Yet, stand up to it they did, and in September 1941, back they came into Kent for the third war-time hopping. One newspaperman who watched them arrive at Tonbridge Station that year reported: "The East Enders showed no outward sign of their ordeals during the winter. They are the good-natured, laughing crowds that were familiar during the hopping festivals of peace days."

But they were certainly fewer than they had been in the peace days, and more and more farmers had to rely upon being able to recruit home pickers, and pickers from the nearby market towns.

Most of the hoppers came by special trains out of London. Previously, although the hoppers' specials had been running still, and carrying thousands of passengers, too, the trend towards individual road transport had begun to make itself felt before the war.

Many of the people who came hopping before the war, too, had begun to arrive by special coach or bus. Many more clubbed together to pay a lorry owner to bring them. That way, they could travel from home to farm in one journey, without having to change trains, or to leave the train and travel by lorry or tractor and trailer or even (still, in those days) horse-drawn wagon to the farm.

There had been gradually creeping into the hopping hordes more and more car-owners; usually people whose work required them to own or drive a car or some other vehicle which could be used to bring the family down to Kent for the hopping, and then taken back to continue the daily work in London until it was time to fetch the family home again.

It all meant that road traffic increased up to the time that war broke out. But then, of course, one of the first shortages was of petrol for civilian use, and so the hoppers flocked back on to the special trains again.

By 1941 the war was real and very earnest, and expectations of an early victory were very muted indeed. Kent was still very much a front-line county, as it was throughout the war, though perhaps never quite so much again as it had been immediately after Dunkirk.

Everyone had settled into a war-time way of life. Air-raid wardens still patrolled the country roads, checking on black-outs, but the need was not so great now. Even in the country-side, people were very well aware of the wisdom of keeping their lights snugly inside their homes at night.

Hopping in 1941 was a comparatively subdued affair, in spite of that local newsman's view of the East Enders' arrival at Tonbridge. The traditional lolly-man, who had used to come round the hop gardens with his trays of sweets for the children, was missing. So were many of the other itinerant traders: the bun-man, the fish merchant, and the others. There could be no camp fires at night and even the pubs were blacked out and much less boisterous than they had been.

There were about 50,000 hoppers in Kent that year. The

brewers allowed extra beer to the hopping villages' inns for the visitors, but the hoppers had to bring their ration cards with them for all their food. They had to register as temporary customers at special shops often set up in barns and other rural buildings, and they were unable to buy food anywhere else.

Not that this was a special hardship, and it was probably true that the hoppers had never, in 300 years of organized casual labour in the hop gardens, been better cared for than they were now.

If food and other necessities were scarce, at least they were reasonably fairly distributed. Because of the difficulties about travelling, many of the bigger camps had resident doctors and nurses; there were crèches for the youngest children on many more farms than there had been pre-war; and many hitherto unknown comforts were provided for the hoppers, too.

This new regard for the welfare and comfort of the hoppers was the result of several different factors. For one thing, there was a completely new regard for the heroism of the Londoners under fire. Almost all of those who came to Kent had been bombed out or had experienced some other incident that set them apart from local country folk. Just to have been in London during the Blitz made them a different kind of people. The Londoners knew it, and made the most of it, naturally, but the people of Kent really did, perhaps for the first time, feel real respect for the hoppers now.

Then, too, there was still a continuing shortage of pickers, and the farmers just couldn't afford disaffected hoppers. They wanted their hoppers to get on with the job and to be ready to come again next year, so they were prepared to pamper them far more than they had ever done before. Then there was the national sense of unity that had, perhaps, never been more widespread throughout all levels of social life. Many of the old traditions, including the rigid class consciousness that existed before the war were falling away under the unifying stresses of war, and the ideas that were to bring the Beveridge Plan and the welfare state to post-war Britain were already taking shape. It would be stupid to pretend that there was no more exploitation, in the hop gardens of Kent any more than anywhere else, but there was less of the old presupposition that exploitation was the natural order of relationships between employers and employed.

The tally in 1941 was, again, about three bushels a shilling. The crop was, again, a splendid one, and farmers generally had all the pickers they needed. Most of the pickers had had their fares into Kent paid for them, and among them were some who had been evacuated out of London, some to places as far away as Blackpool, who came south for the hopping just as they always had done.

On some farms, the soldiers were back on hopping details, helping to make up the numbers of hoppers needed, and a new feature of rural life which was becoming well established by now was the Land Army girls in their green sweaters and khaki-coloured overalls, doing work that had hitherto always been done by men, that of binmen and pole-pullers.

Every farm had its own ARP scheme, and the field-side trenches were now more or less permanent features of the countryside. Many farms encouraged hoppers to join local savings groups for the hopping period, and to buy national savings stamps with part of their earnings. This, too, was something quite novel for old-time pickers, to whom the idea of saving any of their hopping money was completely foreign.

As the war grew older, so, it seemed, did the hoppers. By 1942 there were hardly any men in the hop gardens at all, except those in uniform who were sent there from nearby army stations on a day-to-day basis, and the older farm-workers who were exempt from service in the forces. At their bins, the hoppers sang "The White Cliffs of Dover" and "Down Mexico Way" and other pop-songs of the day, but there was not the old pre-war holiday spirit any more. In many of the hop gardens half or more than half of all the hoppers had been bombed out of their homes at least once, some many times. Almost all of them had someone dear to them to worry about.

Nevertheless, the villages organized dances in the village halls, and they were always well patronized by the younger hoppers. The hoppers' traditional appreciation of the local pubs did not wane, either, and the pubs were still as crowded as ever in the evenings and at weekends. Ration books were by now a normal part of everyday life, but people were now organized to make the best of things and voluntarily-manned mobile canteens were to be seen visiting the hop gardens in 1942 selling tea and three-penny meat pies provided by village

"There they are!" Hoppers took cover among the standing bines when the 1940 Battle of Britain dog-fights went on overhead

Soldiers joined the Land Girls on an East Kent farm to help get the hops picked during the labour shortages of the early years of the Second World War

Using a pole to free bines from hop-poles called for a strong arm and a certain knack

groups. Paddock Wood alone supplied 10,000 of these pies in one week.

Although the hoppers had their own air-raid-precaution organizations, they were under the over-all supervision of the local village organizations. Extra wardens were needed to patrol the rural sectors and these became known as Hoppers' Wardens, whose sectors specifically included groups of hoppers' huts.

The age old problem for farmers of 'scrumping' and other pilfering, far from diminishing during the war years, got worse and the hop-pickers played their part in that trend. With up to 20,000 strangers in a district for three weeks of the year, proper policing of the orchards and gardens was almost impossible at the best of times. Blackout restrictions and man shortages made it more difficult than it had ever been before, especially in a time of shortages and when individual initiative and enterprise were highly regarded and even rigorously taught. Fines of three pounds for stealing apples were little more than tokens of disapproval, and were treated as such.

In East Kent, Mid-Kent and the Weald there were reports of a good crop of hops in 1942, brought to good size and quantity by long spells of dry growing weather with average yields of fifteen hundredweight to the acre. Pickers were paid about sixpence a bushel and very little time was lost during the picking because of bad weather.

It is a little surprising to recall that, all through the war years the numbers of hoppers in Kent stayed very close to pre-war figures overall. In 1943 it was estimated that there were more than 40,000 of them in the Paddock Wood, Staplehurst, Marden, Headcorn and Pluckley alone (these are mostly Wealden areas). There were about 5,000 pickers at work around the Faversham area, most of them local or, at any rate, from other parts of Kent, and perhaps 2,000, virtually all home-dwellers, around Sittingbourne.

This year brought a new feature to the hop gardens. Civil Defence had by now equipped its units with mobile gas cleansing vans—vans in which victims of gas attacks could have been showered to clean off the effects of the gas. They had never had to be used in earnest, but now they were brought to the hop gardens where the authorities were able to achieve the dual effect of testing the equipment and keeping it in efficient working order while, at the same time, providing the hoppers with very much welcomed hot showers.

The same year also brought a less welcome feature to the hop gardens in the form of a serious increase in the hop wilt disease, verticillium. At least 130 Mid-Kent farms reported the disease among their hops and there were serious losses to the crops as a result.

The disease itself was nothing new. It was first recognized in Kentish hop gardens in 1924, when it was discovered at Penshurst but, although it had been a constant threat to the growers after that, it was fairly well contained for the most part until the 1943 outbreak, which came as something of a blow to the farmers.

The season's picking tally was about two and a half bushels a shilling, and reckoning that a good picker could pick about forty bushels a day, takings of about fifteen shillings a day were a reasonable expectation for many pickers. As the war went on, more and more of the old regulars were prevented from coming because they had become involved in war work of one kind or another. One elderly lady told a newspaper reporter in 1943 that she and her two daughters had been coming hopping for many years and usually reckoned to take home about twenty-five pounds between them but that year she had only one of her daughters with her. The other had been prevented from coming by her war work, and had had to be left at home. The old lady said she was crying at the thought of missing her annual hopping holiday when they said goodbye to her.

The majority of pickers from London travelled by train again this year, and some even reverted to the old method of travelling by horse and cart all the way to Kent. The lorries that once so dramatically increased the Kent traffic at hopping time, together with taxis and private cars, were now things of the past, yet there is usually a way round the greatest difficulties, if the desire to find it is great enough, and in the Paddock Wood area, particularly, many of the Londoners arrived by pony and trap, loaded with all the usual items of household comforts intended to make their stay as home-from-homey as possible.

Careful preparations had to be made to feed the hoppers and farms were asked to set aside buildings to be used as shops, complete with a barrier to prevent 'rushing'. This, of course, had been done in previous years, too, and was becoming a feature of war-time hopping. Pickers over fourteen years old were entitled to the agricultural cheese ration, which was greater than that for ordinary civilians. Catering for the visitors, with their ration books and allowances continued to pose problems for the villages, but by this time life was well organized on a war basis and the hop gardens were served by milk bars and coffee stalls, there was hot water for most of them, special facilities for making sure they all had their proper rations from the grocers' and butchers' shops, and, generally, the shortages, although inconvenient, were not serious.

Many of the milk bars, coffee stalls and special shops set up on the farms themselves, were staffed by Land Army girls who had formerly been shop assistants.

The war brought back the horses to the farms that had, before 1939, been mechanizing and on many—probably most—of the Kentish hop farms, the full pokes were tossed into horse-drawn wagons to be taken to the oast for drying, just as they had been a century and more ago. Horses and carts met the hoppers at the railway stations, too, much to the delight of many of the older hoppers who remembered, and cherished the memories, of days when there was no other kind of transport for them anyway.

In 1944, the hop crop was poor, both in quantity and in quality. Generally, the tally was about sixpence a bushel, although in some places it was as high as tenpence. There were still demands for a higher tally on some farms, and when they were not met, or the new offers still did not come up to the pickers' expectations, quite a lot of them packed up and went home before picking was done. This meant that those who remained had to pick more, and picking was going on until dusk and even beyond in some places, without the usual whistles or horns being blown to end picking at any one time. On many farms, high winds had brought down a lot of the bines before picking was due to begin and this, of course, meant that the hops were soft and more difficult to pick.

What with one thing and another, hopping 1944 was one of those hoppings—memorable for all the wrong reasons. Even the wasps, always a significant hazard for hoppers, were more numerous than usual and the voluntary dispensaries treated an exceptional number of stings, especially among the younger hoppers. Of course, this was also the year of the doodlebugs—those pilotless flying bombs also known as V1s. Although they were launched mainly against London, in fact most of them actually exploded in Kent. Kent claimed 2,400 flying bombs landed within the county—100 more than London had.

The open Kentish countryside was under the flight path of a high proportion of the V1s, and made a very good dumping ground for those that could be brought down there before they reached London. The chances of an explosion causing loss of life in Kent were nowhere near as great as if the bombs had reached London and so Kent once more laced its skies, this time with barrage balloons with which the buzz-bombs could collide, upsetting their navigational gyroscopes and causing them to crash. Anti-aircraft barrages were put up against the flying bombs, too, and some of them were brought down in that way, or exploded in the sky by direct hits. More than a few were grounded by a particularly daring manœuvre of the RAF fighter pilots, who intercepted the rocket-powered bombs, flew alongside them until their wings were under the short, stubby wings of the bombs, and then with a waggle of the wings tipped the bomb, again upsetting the navigational instruments and cutting out the doodlebug engines so that the trespassers dived into the fields below. It was hoped that they would explode more or less harmlessly in open countryside. The RAF pilots became quite expert at this particular manœuvre, sometimes getting into position beside the bomb and staying there for some time in order to get well clear of a town or village before scooping the bomb out of the sky and dropping it into fields or woods.

The hoppers organized their own buzz-bomb spotters to give warning when one looked as though it might fall nearby. In many ways it was easier than in the 1940 days of the Battle of Britain dog-fights. The distinctive rasping flight noise of the V1s could be picked out fairly easily from some way off and spotters could make visual contact quite easily on a clear day, helped by the tell-tale flare of the rocket motor. They flew comparatively low, too, and came singly, for the most part, so that once one was heard it was only a matter of listening and watching for the engine to cut out. When that happened, there was a few moments' delay during which a whistle warned the hoppers to take cover—which usually meant dropping to the ground in the hope that the bomb would not fall too near: there was no time to run for more substantial cover.

There were serious doubts about whether or not there were going to be enough pickers in 1944. There was a Government appeal to all who had been in the habit of going to the hop gardens for the picking season, and who were still able to do so, to go again because their help was more urgently needed than ever. They were urged to try to recruit friends and neighbours to go with them, too. Youth organizations were mobilized into hop-picking volunteers and housed in military camps for the season. And so, one way and another, a local newspaper was able to open the annual hop-picking story: "Britain will get its V-day beer, and some to spare!"

Still, it was not a good year, either for the hops or the hoppers. War fatigue was making itself felt in earnest now and people were getting tired of being called upon for that little extra effort. Anyway, the war was almost over. It was clear that it was going to be won by the allies; it just remained to see how much longer the wretched business was going to take and what new nastiness Hitler and the Germans were going to be able to dream up to throw into the last desperate stages of resistance to the inevitable.

And, although by hopping 1945, the war was finally over, the problems were not. Far from it. Immediately before the hopping season began, there were alarms about a serious shortage of food in the hop gardens and criticisms that tradespeople were not being released from the Forces in time to cope with the annual influx of hoppers. The biggest problem was over the supply of bread, and although the shops were set up in the hop gardens as usual, there were real fears this year that they would have nothing to sell, and that the hoppers would go without essential supplies.

With people desperately anxious to 'get back to normal' as quickly as possible, there were more hoppers than there had been in the last few years in 1945. The evacuated Londoners were returning home, and they all seemed to want to renew their interrupted acquaintance with the Kentish hop gardens. Going hopping down in Kent again was part of that process of getting back to normal for them. At the end of August, fourteen trains full of hoppers arrived at Paddock Wood alone, and soldiers and Italian prisoners of war were drafted to the station to help the depleted railway staff to cope with the masses of baggage they brought with them. In spite of the numbers, the station master at Paddock Wood reported afterwards that the crowds were quieter than they had been in the old days, making none of the old demands, backed up by strong language and what he called 'fuss over nothing'.

It was not that the war had broken their spirits. It was that they had learned to put up with things. They had experienced worse than anything they expected to experience ever again, and they found it easier than they had in the past to shrug off mere inconveniences and make the best of things.

After all, things were not too bad in the hop gardens in 1945, in spite of the forecasts. The threatened food shortage did not materialize with the severity that had been predicted. There were some difficult moments at the beginning of the season, but then local tradesmen rallied and there was enough bread and basic essentials for all, even though some of the less essential items were on short rations. There was enough labour for the growers' needs, and enough food for the pickers' needs—especially when Londoners' relatives began to arrive at the weekend with food parcels, having heard about the expected shortages. In fact, all in all, the hoppers probably did better in 1945 than they ever had before and were certainly better provided for in the hop gardens than many housewives and their families were in the towns. The villagers benefited, too. There were extra rations in the local shops, and it was not only the hoppers who shared them. The hop crop was very good that year; though small, the hops were easy to pick and with a tally of about two bushels a shilling, many of the old hands did pretty well.

EIGHT

The Post-War Years

When the blest time of harvest is all o'er
And Ceres bears her golden sheaf no more,
The wealthy farmer joyful locks his barn,
Then looks with pleasure on the smiling gar'n,
Where clustering hops, of a bright silver hue,
Entwine each pole, present a pleasing view.
In the meantime, it is the good man's care
T' provide his pickers, and the bins repair;
The oast-house too is fitted up for drying,
His coals are stow'd, and bagging want no buying.
All things are ready, the hops are ripe,
He strides his nag, he rides, and thus invites:
"Come, my good dames," says he, "tomorrow rise
As soon as Sol illumes yon eastern skies,
My ripened hops again require your aid,
I'll use you well, and well you shall be paid. . . ."

Part of "The Hop Garden" by C. Rofe

In Kent today, people tend to speak of hopping as though it were ended: as though hops were not being picked any more. In fact, of course, hop-growing is still a major feature of Kent's very considerable farming industry and as hops are still being grown, they must be picked.

So, although it is a quite different kind of hopping that goes on in Kent today from that which the county knew right up to the middle of the twentieth century, it still goes on. The difference is that today, instead of the bustling business of hundreds of people at work in each of hundreds of hop gardens, almost all the hops are now picked by machines. What human hop-pickers remain are mostly students, and women living quite near the farms and the hop gardens.

The hoppers of fifty years ago still make nostalgic pilgrimages to the Kent hop gardens, usually in coaches provided by pensioners' organizations in London and the North Kent towns, but the old-style hopping they remember, for the most part fondly, is gone, almost certainly for ever.

Students do a good deal of the work that has replaced the old-style hopping now. They drive the tractors, cut the bines, and sort the hops when they come off the picking machines. In many ways the students have found in the new-style hopping the same sort of reliable annual holiday with pay that the old Londoners used to look forward to. Hopping provides a regular and certain vacation job on an annual contract basis for many of the students today, and this suits both grower and student very well, since the one can thus be assured of his vacation job, and the other is assured of labour at the time he needs it most.

From the point of view of farm management, of course, the absence of the old hordes of immigrant workers is no lamentable thing. Nor have the villagers adjusted to the change without a large measure of relief. The Londoners and other seasonal guests of the countryside were welcomed not entirely without reservations.

Progress is usually double-edged.

The first attempts to make hop-picking mechanized were made in America way back in the 1890s. The earliest machines were not, in fact, particularly successful. They battered the hops pretty badly, failed utterly to distinguish between hops and leaves, and generally tended to deliver an unacceptable sample. Even so, there were plenty of forward looking farmers about, even then, ready to forecast that mechanical hopping was the coming thing.

There were plenty of pickers, too, to protest that such a thing must never happen!

In fact, it was nearly twenty years after those first nineteenth-century experiments that hop-picking machines started work in earnest. That was in 1905, still in America, when the machines picked part of the season's crop there. These machines were an improvement upon the earlier models, but still left a lot to be desired, and represented no immediate threat to the traditional hand-picking methods.

After these early experiments there was very little advance made in mechanical hop-picking techniques until 1918, when some new ideas were introduced, and more machines began to come into use.

Inevitably, as it now seems, once it had been demonstrated that machine picking of hops was feasible, the machines were improved upon more and more. They were expensive, but on the other hand they replaced hundreds of pickers, and pickers were beginning to become more and more expensive themselves.

The first American hop-picking machine to be imported into Britain arrived in 1922. It was a flop. It simply was not designed to pick English varieties of hops. It was not until the early 1930s that further experiments led to the development of a machine which was designed to comb the hops gently off the bines, and which was very much more successful than any of its forerunners had been.

Suddenly, now, the whole future of hop-picking changed. Even the sceptics began to acknowledge that the days of hand-picking of hops might, indeed, be drawing to an end.

All sorts of innovations and improvements were made to machine designs, and gradually they began to be acceptably efficient and more able to rival experienced hand-pickers. A series of cleaning devices were tried, to clean the picked hops of leaves and give a better sample, and one that would bear comparison with those which the hop factors had come to expect from hand-picked hops.

The first demonstration of a hop-picking machine on a Kentish farm was given at Smugley Farm, Goudhurst, in 1937. The inventors were quick to reassure everyone that the machine was not intended to put pickers out of work, but to offset a shortage of pickers. Nevertheless, the pickers were not altogether reassured, and there were strikes against the machines, some of which lasted a whole day or more at a time.

The machine could pick in twelve hours at a rate equal to that of two average human pickers working for thirty-five hours but it had to be staffed by two men and between six and ten women, and serviced by four to six men working in the hop garden, cutting and carting the bines. Even so, this compared with about 160 or 200 pickers needed for the same acreage of hand picking. It was a sign of the times—a first hint of things to come, although the machines were still very far from perfect. They were very expensive, of course, and they still tended to damage the hops. Nor could they distinguish between hops and leaves very satisfactorily, so that their harvesting had to be picked over—cleaned—by human hands but, even before the Second World War, it was already inevitable that machines would one day replace human fingers in the Kentish hop gardens.

One thing that helped to delay the spread of machines was their cost which was, even in 1937, about £5,000. Many farmers were reluctant to invest that kind of capital before it became absolutely necessary and, labour shortages or no, even during the war, most farmers with moderate acreages of hops were still able to recruit enough pickers to meet their needs.

In America, the spread of machines accelerated considerably in the 1940s, once the war brought its shortages of pickers to the hop farms there. Most of the hop-picking machines in use today were built on principles developed in America during the war. By 1950, more than 85 per cent of all American hops were being picked by machines. In England it was less than 10 per cent, but interest in the machines was growing here, too.

For one thing, the old hoppers no longer made up as large a proportion of the total coming to Kent from London every year. The war had begun the change, and the very much improved social conditions, as well as the new full employment, that followed the end of the war hurried the process along. One of the results of this change was that the quality of hand-picked samples declined. Farmers reported that the hops were not being picked as cleanly as they once were, and there were more bunches—something that, before the war, measurers would have refused to take out of the bins.

Growers found it impossible to make demands upon their pickers as they once had, and when faced with an off-hand take-it-or-leave-it attitude they found that they had to take it if they wanted to stay in business at all.

It all meant that the machines, with all their imperfections—and they were becoming fewer all the time—were able to compete on more and more equal terms with hand pickers. Farmers were forced into paying higher tallies, just to get people to come hopping at all. Gradually, the hand pickers lost the advantages they once had over the machines.

As it became more and more difficult for the farmers to get the numbers of people they needed to pick their hops, more of the bigger farmers installed machines. As word got around among the pickers that the machines were taking over, fewer people came forward to pick hops, and so soon even the smaller farmers were having to buy their own machines, too.

By the late 1950s the labour shortage was acute, and in a very short time indeed, the revolution that mechanized hop-picking was complete. There were about thirty machines of three basic types here by then, and in 1954 about 7 per cent of all the hops in the whole of South-East England were picked by machines. By 1958, the figure was about 48 per cent. From that point, there was no going back.

In London, the old East End communities were scattered into the rehousing estates and new towns. Conditions generally were vastly improved, both living and working conditions, and the old incentives to travel to Kent to work all through the one holiday of the year were gone. School holidays were no longer arranged specially to coincide with the hop-picking, whenever it might be, either. What hoppers did still come, more and more often arrived in their private cars, and it began to be necessary for farmers to provide accommodation for the cars as well as for the pickers.

The old hopper huts, for all they were beginning to be called chalets now, had little appeal for people who lived in modern homes with all the conveniences.

In 1954, in his report to the Hop-Pickers' Medical Treatment Board, the Honorary Medical Adviser Dr J. H. Hazledene commented on the "very noticeable" scarcity of pickers on several farms and wondered if University students, who had for so long by then helped to staff the hop gardens' medical missions, might one day be drawn to the gardens not as missioners but as pickers, "seeking remuneration for their own maintenance during vacation". It proved to be a prophetic conjecture. Within ten years that was exactly what was happening, and today most of the seasonal labour in the hop gardens is provided by vacationing students.

In 1956, in the Weald, tallies were around tenpence or elevenpence a bushel, although a strike at Staplehurst pushed the price up to one and threepence for a patch of particularly small hops. A man, his wife, and three teenage children, all used to picking and more or less expert at it, could earn, perhaps, a hundred and fifty pounds in the three weeks: say ten pounds a week each. It was fair money, but it was not enough to lure the former thousands to give up the annual holiday that was by now universal.

As well as machines for picking the hops, there were others including, in 1955, a prototype drying machine, introduced into this country near Marden for the first time. It was an adaptation of a Bavarian herb-dryer and, linked with a hop-picking machine, it meant hops could go from bine to pockets in one four-hour process, with the help of just two men in addition to the hop-picking machine tenders.

A Kent newspaper reporter described 1957 as "probably one of the rowdiest hop-picking seasons ever".

"Brawls at closing time have become commonplace this year," his report told readers. "Officialdom is showing an apparent complete disinterest in some of the most degrading scenes ever witnessed in this area."

Well, even allowing for a certain journalistic licence, and, perhaps, personal ignorance about some of the scenes in the hopping villages in the 'good old days' of hopping down in Kent, clearly all was not well.

Among the hoppers who still came, there was a new element now. The growers no longer exerted the same control of their pickers with implied threats of blacklisting them for future years if they did not behave. Many of the pickers felt they were doing the farmers a favour by coming to pick his hops, rather than vice versa as had once been the case. They were glad of the extra money, but they didn't need it desperately enough to toe any lines they did not want to.

In 1960 the last of the hoppers' specials—those special trains which for almost 100 years had spilled thousands of Londoners into Kent every autumn and which, even as late as 1945 had still brought 30,000 of them in 60 trains—ended the century-old tradition.

This was the big test year for the machines. If they had not met the challenge of the diminishing numbers of hoppers, farmers might have had to re-think the future of the industry. As it was, the machines proved equal to the task. Almost three-quarters of Kent's total hop acreage was picked by machines in 1960. At the most celebrated of all the Kent hop farms, Whitbreads' farm at Beltring, harvested about equal parts of its total crop by machine and by 1,500 pickers.

Although the cost of installing a hop-picking machine was by now about £10,000, most farmers had one.

By 1962 less than 3,000 hoppers called upon the services of the various voluntary bodies active in the hop gardens and only six of the original ten missions were still active by this time.

The annual report to the Hop Pickers' Medical Treatment Board noted: "The steady encroachment of mechanization has been the most telling cause of the changed picture in Kent, but the Welfare State and relative holiday affluence of former East End hopping families have turned the new generation's feet not so much into the Garden of England as through it, *en route* by road and rail for a continental August holiday, money to spend—no need to earn during the hop picking season."

No less than three-quarters of Kent's hop quota was picked by machines that year.

In 1968, farmers were paying remaining hand-pickers one and twopence a bushel. Their machines were doing the same job at a cost of about fivepence a bushel—less than half.

At the Beltring farm, last-night celebrations traditions were continued as enthusiastically as ever, but observers who remembered the 'good old days' recognized subtle differences. There was a band, and dancing, a children's fancy dress competition and a yard-of-ale drinking contest, with a silver tankard prize presented by the farm. There were talent contests, too, and no shortage of entertainers of all ages. It was all very much in the old traditions—and yet . . .

It was, in fact, the final wake for old-style hopping down in Kent. Later that same year, winter brought flooding to the Medway valley and Whitbreads decided its Beltring hop gardens should go completely mechanical the following year, after the farm's hopper huts were submerged beneath three or four feet of water.

So, in 1969, the Whitbread gardens at Beltring, last stronghold of hand-picking in Kent, went mechanical. Today's Kentish hop gardens resound not to the buzz of chatter around the bins or the sound of songs begun at one bin and quickly taken up at a hundred others, but to the throb of tractor engines. The smell of tractor fuel submerges the unique but subtler scent of the hops themselves. Instead of armies of cheerful pickers steadily replacing the green symmetry of growing hops with a ground-covering patchwork of bright-coloured clothing a handful of people, most of them students, drive the tractors and bring down the bines and another handful of people, most of them local women, service the hop-picking machines in their asbestos sheds.

Hop-picking today begins with specially selected loading parties going round the hop gardens cutting down the bines and loading them into trailers. The bines are then taken by tractor to the hop-picking machines, which are usually under cover, although there are mobile machines that can work among the hops.

The machines look like very complicated tangles of machinery, although the principles on which they work are really fairly simple.

Essentially, the machines take the bines which are fed to them, and strip them of their foliage, leaving only the fragile hop cones to be fed by conveyor belt (or taken by tractor) to the oast house kilns where they are dried by fan-circulated air. On some farms, the hops never go to an oast house, but are passed straight from the picking machine on to a drying machine which readies them for pocketing and marketing.

There are three main types of machines: the horizontal, the vertical, and the mobile or portable. The mobile machines work on exactly the same general principles as the stationary models, but they are smaller so that they can be driven among the hops.

The terms horizontal and vertical refer to the position taken by the bines as they travel through the picking mechanism. In the horizontal machines the bines are clipped to graspers mounted on bars which are pulled through the circuit of the machine by bine conveyors. In the vertical models, the bines are fixed to hooks or clipped to graspers mounted singly directly to the bine conveyor chain.

In both cases, the bines pass through banks of wire picking fingers mounted on cylindrical drums or on caterpillars. The caterpillars are assemblies of endless driving chains working on sprockets which pull the picking fingers round. In both systems, the hops and leaves and some of the branches are all detached from the bines. Inevitably, some of the hops get broken by the flailing action of the picking fingers.

Another part of the machine separates the leaf and stalk, leaving the lighter-weight hops to pass out to where they are sorted over by hand to take out all the badly broken hops, leaves and bits of branches.

The plucking fingers are of two basic types. One is a series of narrow pieces of flat metal with bevelled edges that have narrow and shallow serrations along their length. They are flexible and spring-mounted, so that they brush over the vertically-hung bines, which are made to swing, pendulum-like, and also to rotate. The plucking fingers then probe into the bines so that the small stalk on which each individual hop grows is caught in one of the serrations and snatched off. If the hops are fairly heavy on a bine, one passage through the plucking finger bars will not strip the bine, and the rest will have to be picked off by an arrangement of rotating shafts with flat steel 'flying fingers' mounted on them.

These then strike the hops off the bines. The picked hops, leaves and stalks drop on to a horizontal wire mesh conveyor which riddles out the loose petals and then passes over an upward draught of air which cleans the hops further, leaving the cleaned hops to be discharged on to a final conveyor belt from which they are bagged or, sometimes, fed directly into the oast kiln or the drying machine.

Some machines have picking fingers that are actually wire loops that lasso the hops on the bines and pull them off as they spin.

The output of the machines varies, but is about 300 pokes (3,000 bushels) in a 12-hour working period. Generally, the English machines are smaller than the American models. A double-plucker-bank machine assembly occupies about 110 feet by 20 feet, but some other machines are less than half this length and narrower, too.

The old oasts remain on many farms, but they, too, have changed. Modern drying methods make tall, cowled drying chambers unnecessary. Modern oasts, too, can dry more hops more quickly than the old ones using the old methods could, so fewer are needed.

Fortunately for the traditionalists who have no wish to picture the Kentish countryside without its round oast kilns and white cowls, many of the redundant oasts have been converted into storage buildings and many more are now very attractive and much sought-after private homes.

Once all the farmers who needed them had their hop-picking

machines, of course, the industry that made them declined. Now there is only one company, in Worcestershire, making them and business is not exactly booming for them. A new hop-picking machine costs about £30,000 now, and most hop growers prefer to patch up the machines they already have rather than buy new ones.

For the moment they can do this by buying the machines of farmers who are leaving hop-growing to others, and cannibalizing the old machines for replacement parts. This process is likely to go on for a while, as more small growers get out, leaving the industry, and the machines, to the dwindling number of major growers. But when all the redundant old machines have been bought and stripped down, the remaining farmers who need machines will have to depend on foreign suppliers.

The whole hop industry generally is in a very uncertain state at present. So many changes are taking place, or are forecast for the near future, that many farmers—even those for whom hops have become a major part of their annual product—are marking time to see what will happen next.

There is a UK surplus of hops and the Common Market countries have traditionally preferred the Continental seedless hops to the almost exclusively British seeded varieties. Although Continental brewers have been known to make their lighter lagers from British-grown hops in times of no alternative, they usually insist that the seeded hops are no good to them, and refuse to buy them.

So that is one of the problems facing British growers at the present time and, indeed, some are bowing to what they regard as the inevitable, and are going over to growing seedless hops in anticipation of the new, much enlarged, European market.

Pessimists—or, perhaps, realists—forecast specially difficult times ahead for the small hop growers; some go so far as to foretell his disappearance altogether before very long, as the already big growers become even bigger, and the small ones abandon hops for crops which are less trouble. Hops have always been a crop for the farmer who could afford the high capital costs and the considerable element of risk involved, and that is still true today.

There has, too, been a long and steady decline in the quantities of hops needed by the brewers. Many breweries have amalgamated to form larger groups, and this has resulted in economies and less waste, of hops and of other ingredients as well.

Then, too, the chemists are able to provide the flavouring and preserving qualities traditionally provided by the hops. The chemical substitutes are more convenient and more reliable than the real thing, and if a few—probably a very few—beer drinkers can still recognize the true taste of hopped beer when they are able to find it, how many of them care enough to affect sales enough to even delay the seemingly inevitable universal adoption of the chemical version?

One of the main reasons for the declining enthusiasm for hop growing among small and comparatively small growers is that historically Britain's hop industry has been cut off from export markets because foreigners have not wanted English hops. Their beer-drinking tastes have developed along different lines from ours, and our hops have not been suitable for Continental brewing. English hops only went abroad in any quantity when there was a global shortage. The Continentals, however, have not suffered similarly; the Germans, particularly, have sold hops to every beer-producing country in the world—even though about half of them are of varieties originally developed in Kent, at Wye College Hop Research Department, in fact.

There will be further reductions in needs as brewers, more and more, turn away from traditional varieties of hops, and it could be that some 30 per cent of present acreages will become surplus to home needs.

It may be that Kent is destined to take a secondary role as a hop-growing region in future, behind the West Midlands, where growers are now showing more eagerness to take up hop quotas than are other Kent farmers. This is partly due, no doubt, to the serious ravages caused among Kent varieties of verticillium wilt and other pests which is not so serious in some other parts of the country and which can only be eradicated or prevented from spreading at considerable cost to growers.

However, while hops are still grown in this country—and there is no real reason, certainly at this stage, to suppose that they will not still be grown here for many years to come yet—there are likely to be hop gardens in Kent, for all the same reasons that established them there in the first place and made Kent the number-one-hopping-county of England.

The aesthetic value of hops growing in Kentish hop gardens is beyond price for the traditionalists and the conservationists, but aesthetics tend to be valued more in outgoings than income where the farmer is concerned, and farming is an industry in which efficiency and hard-headed economics are vital. Hops, like all other crops, must be harvested to provide the grower with a cash return on his investment in them, so while hops are still grown in Kent, there will still be hopping down in Kent but quite what that phrase will mean in the future, no-one can be sure. It is only very safe to forecast that it will never again mean the same as it did in the heady hopping hey-day of a century ago.

NINE

The Hoppers' Specials

Some travellers were got together,
Talking and yarning at the weather,
At last, to change their conversation,
They talked of sights in every nation.
One who had seen great Teneriffe,
Whose wond'ring size exceeds belief,
Swore "From the top, an house's height
Was not so big as any mite;
Nay that the ships upon the seas
Did scarcely look as large as bees:
A thousand miles at least, at once
You might with eyes take in your sconce."
A Kentish man of ready wit,
On which sometimes perhaps you'll hit,
Said that a Hop-ground was confest
The finest prospect and the best:
"For there," says he, "upon my soul,
"You'll see at once from pole to pole!"

Jan Ploughshare
(Quoted from Kentish Chronicle *1802)*

The coming of the railways brought enormous changes everywhere. In Kent, they opened up the Thanet and south coast resorts, and practically created towns like Margate, Ramsgate and Broadstairs. They were almost wholly responsible for the Romney Marsh resorts of Greatstone and Littlestone and they completely changed the whole character of towns like Tonbridge and Ashford.

Kent has always been part of London's countryside. As London has pushed its own boundaries further and further out from the centre, that countryside has moved out with them. At first, of course, it was the wealthy who drove in their carriages out into the countryside, which then meant places like Woolwich and Greenwich and the fringes of Blackheath.

To be sure, the poor people of the East End of London trekked east and south into Kent in search of work in the fields and the orchards and the hop gardens, but theirs was a very different kind of pilgrimage, differently motivated and very differently carried out.

It took days for the eighteenth- and early nineteenth-century hoppers to journey into Kent and find a farmer in need of their labour. A few came in horse-drawn vehicles of one kind or another, but thousands more, each year, walked. It was, after all, only some thirty miles from London to, say, Tonbridge, and well within walking distance.

Many of them found the work they sought no more than ten or fifteen miles from home, in the hop gardens of those parts of Kent which have only quite recently been engulfed by London's sprawl—Bromley, Bexley, Wickham and Beckenham, for example, but for most of them it meant the long walk to the Weald, sleeping rough overnight, begging or foraging for food on the way, and taking a chance that the weather would be reasonably good.

The railways changed all that. They pushed into Kent in the 1840s. First the South-Eastern Railway extended the London–Redhill line through to Tonbridge, Ashford, Folkestone and Dover between 1841 and 1844, opening the line to traffic in February 1844.

In the same year, link lines from Paddock Wood and Maidstone, and Tonbridge to Tunbridge Wells were opened. At once, the Weald of Kent was brought to within a few hours, journeying time of the centre of London.

In 1846, Ashford was linked with Canterbury, Ramsgate and Margate, and although this line was always used more by the new sea-side day-trippers, it put the East Kent hop gardens within the network of lines that now laced all Kent. In less than ten years, the whole of the Weald and much of East Kent too was opened up by the railways, with the hub of the Kent system at Ashford.

North-West Kent plugged into the railway grid in 1856 when a line linking Maidstone with Strood was opened, to complete the line which had been opened in 1849 between London and Strood. The North Kent coast had to wait for the East Kent Company to build the line from Strood via Rochester, Chatham, Sittingbourne to Faversham in 1858, but two years later the line was extended through to Canterbury and then, at last, East Kent was well and truly on the railway map.

Branch lines connected with Whitstable and Herne Bay in 1861 and, in the same year, the line from Canterbury was extended to Dover, which now had a more direct link with London than the South-Eastern's line via Ashford and Folkestone provided. In 1863, the North Kent route was carried through to Margate and Ramsgate.

At the end of the century, the South-Eastern and the later-named London, Chatham and Dover Railways combined to form the S.E. and Chatham Railway Company, and between 1862 and 1895 new lines were laid to link West Kent towns and villages.

The Reverend R. Shindler, in his book *Among the Hop-pickers in the Weald of Kent*, wrote of the hordes of migrant hoppers who arrived by train in Kent: "Nothing more squalid can be imagined—they often appear to be literally one seething mass of rags and tatters."

It was like an annual gold rush. Some of the hoppers had no possessions at all. Some brought with them most or all that they could call their own: pots, pans, even the family mattress. The railway companies were quick to recognize the passenger potential of these thousands of hoppers and by the mid-1850s regular hoppers' specials were bringing the Londoners to Kent for the annual hopping.

The first of these specials were provided by the S.E. Railway Company. Although intended to carry the hoppers, the trains were not made up with passenger coaches, but of cattle trucks, into which the cheap-fare East Enders were packed and left to make themselves as comfortable as they could on such boxes and stools and mattresses as they chose to carry with them.

The S.E. Railway Company carried hoppers from London to any station between Tonbridge and Strood for two and sixpence, starting at 6 am on Sunday. By 4 am, Bricklayers' Arms Goods Station was seething with several thousand Londoners all waiting to take advantage of the cheap railway fare into Kent.

Children in arms were allowed to travel free—and there are stories of grown children pushed into handcarts or even carried, swaddled in blankets to disguise their real size, being claimed as 'children in arms' by their frugal parents.

Some who could not reasonably hope to get away with that particular ruse were smuggled into compartments and hidden under seats and the station masters at arrival stations had the job of collecting unpaid fares from such human contraband. Those who couldn't pay in cash, could be required to pay in kind, and the station master at Paddock Wood was known to confiscate the shoes of bilkers.

Most years, the hoppers' specials carried more passengers back to London at the end of the hopping than were carried down for the start of it. Hoppers who could not afford even the reduced fares before hopping, could often decide it was worth spending part of their earnings on the luxury of a train journey back home again.

In 1865, special trains from London, Woolwich and Gravesend carried 11,090 fare-paying passengers into the Kentish hop-garden areas. At the end of the season, about 12,000 returned home by train. In 1876, about 36,000 people left London to go hopping in Kent. The train fare from London to Paddock Wood was about two shillings but many still could not afford it and walked. Some came by boat down the Thames to Gravesend and then walked the rest of the way from there.

But many did travel by train, often leaving London during the night and arriving in Kent very early in the morning, where they were virtually stranded, without bed, food or drink, until it was daylight, or until the farmers' horse-drawn wagons arrived to collect them.

By 1880, the number of hoppers' specials passengers carried by the two Kentish railway companies was 18,902 down, 22,607 up. But, not to make too much of this upward trend, traffic varied considerably from year to year, largely according to the changing needs of the farmers for pickers, and in 1881 the passengers down numbered only 14,951, rising to 17,589 for the return trip.

The figures, of course, are no sort of guide to the numbers of Londoners who went hopping in any one year. The hoppers' specials only ran at particular times of the day, usually starting early in the morning. Not everyone was able or willing to use them, even for the considerable reduction in fares. So some travelled by ordinary service trains as full fare-paying passengers. Right into the twentieth century, still many made their own way to the hop gardens, less often then on foot, but many by horse and cart, often in parties organized to combine to pay the owner for carrying them.

In the early days of the hoppers' specials, it was quite common for many of the hoppers to give up their homes at the end of a week and simply leave for the hopping, intending to find a new home when they returned. In the meantime, the week ended and the rent paid (or not, as the case might be), many of the Kent-bound Londoners were homeless until they found a lodging for the hop-picking period in one or other of the Kentish farms.

Because of this practice, the railway companies introduced night-trains, so that the homeless hoppers could spend the night at the London railway station and in the train, arriving in Kent early next morning ready to look for work and a fortnight's or three weeks' lodgings. This practice was opposed by many of the voluntary welfare workers in the hop gardens because it stranded passengers, often with a high percentage of very young children among them, at village stations with hours to wait before they could go to their farms. Sometimes the station authorities would not allow them to remain on the platforms, and the hoppers had to tramp through the country lanes until they found a village green or a field where they could settle down for however many hours' waiting remained.

Until the railway companies were persuaded to end this night-riding practice, some of the hoppers' missionaries made a practice of providing canteens on or near the stations and in the villages, where the hoppers could at least get a hot drink and something to eat while they waited.

For the most part, the cheap fares applied from specified London stations to specified Kent districts. Some farmers, whose farms were beyond the range of the specials and their concessionary fares, paid the difference between the concessionary fare and the full fare for the pickers they employed.

The hoppers' specials were a feature of railway and Kent life, once a year, for about a hundred years. They provided railway staff with all sorts of problems, but also they added something to the lives and experience of everyone concerned with them. They ran from the last week in August for about two weeks, carrying not only the hoppers themselves, but also the flocks of weekend visitors, relations and friends, to the hoppers in their encampments.

In the early days there was a good deal of rowdyism, and the policemen who were always on duty at the destination railway stations were not by any means on an easy number. The arrival of a hoppers' special at a village station was always a noisy occasion, attended by a good deal of shouting, swearing, children crying, and usually some song-singing during waiting periods. Children became separated from parents and had to be reunited, usually with the help of screams from the child and shouts from the parent.

Luggage became mixed up, one family's possessions being claimed by another family, and reclaimed by the rightful owners, all, it seemed, at maximum lung-power. People wandered away from their own parties and ended up in the carts waiting patiently to lumber off back to the farms, and there would be further delays while these confusions were sorted out.

For the most part, it was all good-humoured enough, although inevitably there were unpleasantnesses now and again—hence the presence of the policemen.

By the mid-1920s, the Southern Railway (as it had by then become) was entraining about 30,000 pickers at London Bridge and New Cross stations, as well as smaller numbers from North Kent, and taking them to towns and villages throughout the county.

Obviously, this sort of once-a-year traffic volume needed a special organization in order to handle it efficiently. A Joint Transport Committee of railway representatives and farmers was set up to help co-ordinate rail transport needs, and the railway company appointed a Liaison Officer for the season, with an office at the headquarters of the National Farmers' Union at Maidstone.

So that the railway company could lay its plans for coping with the hopping travellers properly, postcards were printed by the company and distributed to hop-growing farmers. As soon as a farmer knew the date when hop-picking would start on his farm, he stamped and addressed the cards to his pickers and sent them all to the Liaison Officer, with a note of when he wanted the pickers to arrive.

From the information provided by the farmers in this way, the Liaison Officer was able to decide how many trains would be needed, and the times at which they would have to run. These details were entered on the cards by him before he posted them to the pickers to whom they were already addressed. He might have to handle up to 2,500 of the postcards on any one day, but the system kept the conveyancing of the pickers under control and created a very good semblance of order out of what could otherwise have been utter chaos.

Because the farmers knew the times when their pickers were due to arrive, they could have wagons waiting for them at the stations to take them on to the farms, and although to wondering locals—especially the children who hung over the station railings to watch the once-a-year spectacle—the whole business of off-loading the trains and loading the wagons to the accompaniment of uninhibited Cockney comments looked and sounded like confusion of a very high order indeed, in fact it was usually accomplished with easy-going efficiency and a great deal of good humour all round.

Sometimes the farmers would come personally to the station to greet their pickers, many of whom were old friends, and there would be warm—almost affectionate—reunions.

Arrivals off the specials began to reach their farms at about 5 am. Well before that, lorries were arriving with hoppers who had made their own way down and hoped to get the best of whatever accommodation was going before the rush started. Some of the regulars had what were regarded by them and by everyone else as 'their own' huts, but in many cases it was a matter of first come, first served.

The farm carts to collect the rail passengers were not usually at the stations much before seven or eight o'clock in the morning.

The early morning travelling was not only convenient for the hoppers, who thus had most of the day in which to get themselves settled in when they arrived at their farms; it was very much to the advantage of the railway companies, too. It meant that most of the specials completed their runs during the 'dead' period of normal passenger traffic and so interfered very little with usual time-tables.

At the London end, the hoppers arrived at London Bridge and, after buying their tickets, were directed to one of three departure stations there: High Level, Low Level, or Central.

Each card posted to a hopper by the company's Liaison Officer bore a letter identifying the train on which the passenger would travel, so that station staff at London Bridge could direct the travellers to whichever platform the appropriate train would leave from. In this way, passengers could be marshalled in trains according to destination.

Throughout the duration of the hop-picking season, the railway companies ran Hop-pickers' Friends trains at weekends, too. Passengers on these trains travelled for the ordinary week-end ticket fare on Saturday and Sunday. Most of them made the return journey on Sunday evening, and sometimes the railways handled anything up to 20,000 of these passengers between 6.30 and 8.30 on a Sunday evening.

Just as hop-picking did not start on exactly the same date on every farm, neither did it end all at once, either. Smaller farms with fewer hops finished before the bigger ones with a greater acreage to be picked, so that the hoppers' trek back to London usually covered a period of up to about three weeks.

The 'up' journeys were as carefully organized as the 'down' ones. As soon as each farmer was able to say when picking could be expected to end on his farm, he could decide the day on which he would pay off his pickers. That day was notified to the railway company's Liaison Officer at Maidstone who, once again, organized the times of the trains. That done, he passed the time-tables on to the farmers, whose job it was to make sure the hoppers knew them. Usually this was done by pinning up handbills on farm buildings, around the hoppers' camps, and anywhere where the times could be studied while the pickers were waiting to be paid off.

On the day of departure from Kent, the hop-pickers would congregate at the appropriate railway station, normally on a weekday afternoon, for on-routing to Paddock Wood which, for the period, became the main marshalling point for all the south and east Weald hoppers.

By the mid-twenties the system was working efficiently enough for the hoppers to be practically guaranteed a journey of not more than about two and a half hours from entraining at the local station to arriving at London Bridge. Time-tables were arranged so that the hoppers arrived home earlier than they had once done, too. Whereas, some years before, it had been normal for quite a lot of the trains to start the homeward journey between 7 and 11 pm, now the last hoppers' special headed for London at 7 pm.

This not only meant the hoppers got home at a more convenient time of day; it also meant that fewer of them were persuaded by the pubs along the way to the stations to stop off, forget all about the meticulous arrangements made for them, miss the last train, and create the problems attendant upon such frequent mishaps the following day.

Railway staffs reckoned it took about twelve minutes to load the luggage of a hundred passengers at country stations. Punctuality was not an East Ender's virtue at the best of times. Farewell celebrations, delays in paying off, lost children—and adults, too, for that matter: these and a hundred and one other reasons for arriving at the station at the very last minute, just as the train was about to steam out, all meant that the departure of a hoppers' special was an event that could strain the good humour of everyone involved to breaking point.

Sometimes there were fights on the platforms; baggage already stowed away was demanded back again because something in it was needed, or because something not needed had not been packed away; for as many reasons as there were families, the confusion was considerable, the noise incredible.

Yet, almost always, the trains did depart at least more or less on time, and for a little while afterwards the comparative quiet on the station and in its immediate neighbourhood was almost profound.

In one year (1925) the railway ran 71 hoppers' specials into Kent, and 75 for the return journey. They carried 34,448 hoppers by special and ordinary trains combined to the hop gardens, and 22,177 back again. That was one year when, apparently, the usual trend for the volume of passenger traffic to be greater on the return trip did not apply.

In addition, during five weekends that season, the trains carried 33,980 Hoppers' Friends into Kent on the Saturdays, 13,854 on the Sundays, and 43,412 back again on the Sunday evenings.

It was a major undertaking for the railway, every year, for just about 100 years. The hoppers' specials were a very significant part indeed of the story of nineteenth- and twentieth-century hopping down in Kent.

Quite apart from scheduling the trains, extra staff had to be brought on duty, particularly at the London end, and arrangements had to be made to accommodate as many as possible of the hoppers inside London Bridge Station, so that they would not interfere with normal road traffic outside. That in itself was no inconsiderable task, considering the sheer numbers of people who arrived there during the few hours before and up to departure times. As soon as one train was away with its quota of hoppers from one of the platforms set aside for the operation, the next lot were allowed through the barriers on to the platform to wait for the next departure, and to ease the congestion in the main concourse and approaches to the station.

Some of the country stations were busy if they played host to twenty or thirty passengers a day through most of the year, yet, suddenly, late in August or early in September, they had to cope with hundreds. Special temporary shelters had to be set up, and special once-a-year procedures brought into effect to cope with the sudden rush.

Even through the war, the specials still ran. In 1940, the railway had to organize special services for evacuated hoppers who came to Kent for the season and then went back, not to their old homes in London, but to new evacuee homes, some of them as far away as Birmingham and Warrington and the West Country.

Perhaps the railways were thanked for their work. Compliments are never as conspicuous among archives as complaints are and certainly there were complaints. One that created quite a substantial body of correspondence was that two of the specials that ran in October 1940 from Marden to Birmingham and from Wateringbury and Headcorn to Warrington respectively carried large numbers of children but no water for them to drink or wash in during the journey. That particular complaint was bandied about between Southern Railway officials and men from the Ministries, for some months. In the end, the railmen admitted that it was a very long journey for children to be waterless, but pleaded special difficulties caused by the war.

From first to last, the hoppers' specials gave Southern Railway a memorable chapter for successive generations of railwaymen, and contributed nobly to the hoppers' lore whenever they reminisced among themselves.

The last hoppers' specials ran in 1960 and when the last train left Kent that year, taking home the last of the season's London pickers, they brought to an end a colourful era that had characterized Kent for longer than anyone could remember.

TEN

The Oasts

> The hop for his profit I thus do exalt,
> It strengtheneth drink, and it favoureth malt;
> And being well-brewed, long-kept it will last,
> And drawing abide, if ye draw not too fast.
>
> *Thomas Tusser, sixteenth century*

The focal point of all hopping activities was, and to some extent still is, the oast house. It is still the drying process, however carried out, that converts the hops from farm produce into a brewing ingredient.

Because, ideally, hops need to be dried quite quickly once they have been picked, it was always from the oast house that the key man in the whole process, the dryer, issued his directions about how many more hops could be picked and when picking should stop for the day.

Modern drying methods have relaxed the benign tyranny of the old-time dryers to a large extent. Many more hops can be dried more quickly now than they once could. But in the hey-day of hopping—particularly throughout the nineteenth and early twentieth centuries—the dryer was a man who wielded a great deal of influence over the pickers, even though there might be no direct contact between the two.

The oast house was a social centre of the hop-picking season, too. On many farms, where the oast was near enough to the gardens, there were parties in the evenings, and especially at weekends, at the oast house, with hoppers and binmen and probably some visitors sitting around on piles of the coarse hop pokes or stacks of the finer jute hop pockets, or on anything else that happened to be available, drinking, singing or yarning, or just soaking up the warmth from the kiln fires and growing more and more sleepy in the heavily aromatic atmosphere as the evening wore on.

At the end of the hop-picking, it was at the oast house that the hoppers often gathered for the traditional end-of-hopping bonfire and feast that always included potatoes baked in the ashes of the fires, washed down with beer and cider from casks tapped for the occasion.

To people whose experience of hopping is limited to what they see as they pass through the county, the term oast house certainly conjures up a mental image of round kilns and dunce-cap roofs topped by white cowls with a flat wooden vane often ridden by a prancing horse. In fact, though, these 'typically Kentish' landmarks are comparative newcomers to the county scene.

When the cultivation of hops first came to England from Flanders in the early sixteenth century, all sorts of improvisations were achieved for the necessary drying of the hops. At first, farmers were advised against using any form of artificial heat, and to rely on the warmth of the sun for a proper drying. But the growth of the industry soon made this too slow and too unreliable. Existing farm buildings were cleared and contrived to provide a drying floor where good ventilation was the first need, since a wholly natural drying process was still regarded as vital.

But once the industry was well established and it was clearly worth the capital outlay to do so, farmers began to build special drying houses modelled on those already familiar on the Continent. These were just rectangular buildings, usually about 20 feet by 10 feet, divided into three rooms or compartments. The middle compartment was the drying floor. At one end was the loading area, where the hops were brought from the hop garden; at the other end, a cooling room into which the hops could be raked or shovelled when the drying process was complete, and where they could be bagged ready for marketing. Thus a continuous through-put was possible, with picking controlled to ensure that hops did not lie too long in the loading room before passing on into the drying floor.

These early oast houses would not be immediately recognizable today to anyone familiar with the later, nineteenth-century oasts, but they served their purpose well enough for a couple of hundred years.

One of the things that encouraged the greater acreages of hops in the eighteenth century was the advances that were made in methods of drying the hops. At the beginning of the century, most oast houses worked on a purely domestic basis, supplying the farmer who grew them, his family, and perhaps their friends and workers. At that time, the hops were spread on a drying floor in a barn set aside or perhaps even built specially for the purpose, and simply left to dry naturally.

Reginald Scot (1574) advised growers to build their own oasts (he called them ostes) and he suggested a little house 18 feet long by 8 feet wide, with three rooms. The middle room, 8 feet square, was for the drying area. On either side, he said, there should be other rooms, one for green hops and one for the dried hops. Thus, the hops could come in from the gardens at one end of the oast house and be processed through to the other.

In the drying chamber, it was recommended that a ceiling should be built 5 feet above the ground floor, made of laths one inch square and arranged a quarter of an inch apart. The oast walls should be 9 feet high. Below the drying floor, there should be a brick furnace, 13 inches wide and deep, 6 feet or 7 feet long, shaped like a long prism with holes in the side to distribute the heat, and filled at one end with the fuel (charcoal or wood). The laths on the drying floor were then covered with a blanket made of horsehair, on which the hops were spread an inch and a half thick, to be dried.

William Harrison (1557) described hops being dried in a garret or upon the floor of an upper chamber, but his advice was never to use heat to dry them, nor to dry them in the sun. He believed the drying loft should be open to the air, so that the hops could dry slowly and naturally, being turned at least once a day for two or three weeks.

When some farmers first began to experiment with heat drying, it was generally regarded as a stupid extravagance. To cut down on this extravagance somewhat, some farmers got together and dried their hops in a common malt oast. It meant that they could dry a greater quantity of hops during the season, and so they could grow more and sell what they did not need for their own use. It was good business, and soon heat drying became the normal process.

Oasts became larger. The now familiar two, three or more kilns, usually at one or even at both ends of the rectangular two-storey barn, pushed up the quantity of hops one farm could cope with, and since the demand was good, farmers constantly experimented with ways of improving their drying capacities.

Because of the need to dry the hops quickly after they were picked, the dryer in his oast was always a key figure in the whole hop-picking process. It was the dryer who sent out word when picking had to stop because the hops were arriving at the oast faster than he was drying them, and it was upon his skill and dedication (hop-drying was a twenty-four-hour-a-day job, and dryers often had to make do with cat-naps throughout the three or four weeks of hop-picking, with perhaps a bit of real sleep on Sunday, if they were lucky) that the success of a season's hop-picking depended to a very large extent.

By the second half of the eighteenth century, when new ideas seemed sometimes to be the best British farm crop of all, the forerunners of the round kiln oasts had begun to appear. These were usually built with a square kiln that had a cowled pyramid for a roof, joined to a barn-like building, sometimes known as the stowage, into which the hops could be loaded before drying, and cooled afterwards, ready to be pressed into pockets and sent to the market.

H. M. Manwaring in his *Treatise on the Cultivation of Hops*, in 1855, set out these instructions for the proper design of an oast house of that period:

> "If the kiln be 12ft. square on the top, it should be 12ft. from the fire, and the staddle should be 6½ft. square, and so proportionately in other dimensions. The fireplace should be about one foot square, and over it a small door or shutter, about 18 ins. square.
>
> "The bed should be of laths or rails, very even, about one inch thick, and the same distance apart, and covered with oast hair. On this cloth, the hops are emptied out of the pockets about six or eight inches thick."

Oast hair was the name given to a blanket of horsehair that was laid on the laths forming the kiln floor. It allowed the warmth from the fire below to circulate through the drying hops without permitting the hops to fall through the floor. Later, some kiln floors were covered with sheets of perforated metal gauze fixed to the laths. When the hops were dry, they were shovelled out on to the ordinary wooden flooring in the

The oast houses at Beltring, illuminated

Hops being pressed into a pocket in an oast house at Hawkhurst

Stoking the furnaces in an oast house

chamber where they were pressed into hop pockets. The shovel used for this unloading of the kilns was a special one, some two feet wide and made of wood or, sometimes, of hop-sacking on a wooden frame, which made it look something like a wide shrimping net.

Incidentally, Manwaring seems to have been wrong when he referred to the hops being emptied out of the pockets. The sacks or bags into which the hops were measured from the bins in the hop gardens, and in which they travelled from the hop gardens to the oasts, were called pokes. They held ten bushels of hops each, and were quite different from the pockets into which the dried hops were pressed ready for marketing which, when filled, were solid tubes of hops, about six feet tall, and weighing about a hundredweight each.

Manwaring advocated drying fires of charcoal, coke or Welsh coal, with brimstone added to the front of the grate. In his day, the brimstone was usually offered into the fire on a flat iron shovel or pan. Later, it was sometimes tossed into the fire by the handful in the form of broken sticks, very like bright yellow seaside rock—without the lettering!—and very hard. He reckoned the drying hops would be ready to be turned after between seven and nine hours, then they would need another two or three hours to finish drying properly. It was the judgement that went into deciding whether the hops were ready for turning after seven hours or if they needed to remain undisturbed for nine, or until some point between the two, that made the dryer such a very important man to his employer.

The square kilns were invariable until early in the nineteenth century. Unfortunately, though, it is not possible to say that all square kiln oast houses remaining today are necessarily older than any round kiln oasts. Towards the end of the nineteenth and into the twentieth century, square kilns regained some of their popularity, although now they were usually larger than the eighteenth-century ones.

The round kiln oasts that so delight the eye all over Kent were built throughout the nineteenth century. The invention of the circular kiln with the revolving cowl is credited to one John Read, whose chief concern, naturally enough, was for greater efficiency in drying hops. Incidentally, though, he bequeathed the hop-growing areas many different examples of their most charming and most characteristic landmarks. Many of them

I

are no longer used for their original purpose. Some have reverted to use as barns and storage accommodation; some have been converted to other uses. Those that do remain in use have, often, been adapted to the changed needs of modern methods so that they serve the farmers better but, sometimes, look like caricatures of the oasts we know and love.

Many of them, all through the Weald of Kent especially, have been converted into private houses of great charm and character (although some of the early conversions were less happy and have endowed the county with one or two quite hideous nondescripts) that will, probably, grace the Kentish landscape for many years to come.

The reason for the distinctively inverted funnel shape of the kiln roofs, originally, was simply to improve the upward flow of warm air from the fires below, and so increase the draught into the fires. The kilns were designed to rely on natural draught for maximum efficiency and round kilns were thought to be more efficient than the older square ones until it was demonstrated, after almost a hundred years, that they were not, and the cheaper and more easily built rectangular kilns came back into favour again.

All through the nineteenth century, the acreage of Kent hops increased rapidly, and many farmers had to decide whether or not to stay in competitive business by enlarging their oasts. Those who did decide to do so usually achieved the desired result by the comparatively cheap and fairly simple method of adding new kilns to existing buildings. It was the way these new kilns were added that created the great variety of grouping that has left us with so many individually unique oast houses in the county today.

Now, oast houses can be fitted with fans to force the draught and make the specialized design of the kilns no longer strictly necessary. Many of the old oasts are still in use, modernized internally with new equipment, but from the outside still looking very much as they always have, but where new oasts have been built they have usually been designed as plain barn-like structures, with louvred vents—really not so very different from the first oasts, three hundred years ago.

The increasing mechanization of the hop industry generally has changed many things inside the oast houses. The old drying methods have been brought up to date and today there are several different processes in use. One involves a continuous belt on to which the hops are loaded and which takes them through a drying chamber and out again, making unnecessary the laborious and skilled work of spreading and turning and finally unloading the hops. Another of the new systems is known as the descending column system, in which green hops are loaded into the top of a tall enclosed column in which they are dried *en route* to the bottom of the column where they are removed, ready for bagging.

The two-tier system divides the drying chamber horizontally so that two floors of hops can be dried at the same time.

They are all ways of speeding up the process which, in the old days, was done with great deliberation and even a certain amount of mysticism. Dryers were the élite of the hop industry. It took a long time to gain the necessary experience to become a properly proficient hop dryer, a great deal of whose skill depended upon the senses of smell and touch and upon a fair measure of sheer instinct.

During the whole of the hop-picking season, dryers often never slept more than an hour or two at a time, always in the oast house, usually on a rough wooden bunk made up (but sometimes built in as a permanent feature) in a recess against one of the kiln walls. They never removed their clothes—not even, in some cases at least, their caps, and certainly not their boots; and there was always a barrel of beer (or cider, or both) set up in the oast, for drying was thirsty work, and the hop dust parched a man's throat.

Although hop-picking time elevated a dryer to an enviable status on most hop farms, the rest of the year he was just one of the farm workers. Traditionally, he earned extra money during his stint in the oast every year, but the rest of the time he would probably be earning just the same as his fellow workers.

In the same way, once the last load of hops had been dried and pressed and made ready to be carted away to the market, the oast houses themselves returned to more mundane duties for the rest of the year. They were usually used for storing other produce—apples, perhaps, or beans, potatoes or other root crops. Some became farm workshops when they were not drying hops.

There was always some work for them to do during the eight

or nine months of the year when they were not doing the work they were specifically designed to do. They were, after all, hefty pieces of capital investment, too valuable to be allowed to lord it over the countryside in haughty idleness between times.

ELEVEN

The Language of the Hoppers

Around the brown board at the farmer's we met,
Where plenty of all we could wish for was set;
His hops were all pick'd, and of corn his barns full,
Man and wife are all jolly, 'twas a sin to be dull.

He, blest with his friends, with his children and spouse,
Gave freely, drank freely, and bid us carouse,
By Jove, we enjoy'd it, as sons of true mirth,
We drank his success in the fruits of the earth.

We drank, sung and danc'd, and told stories of fun,
Ne'er heeded old Time, nor his sands how they run:
'Twas the farmer's good will we of joy should be full—
We resolv'd to be so, and hang all that were dull.

Part of "The Hop Supper"
quoted in the Kentish Garland, *1881*

Naturally enough, during the course of several hundred years of hop-picking, there grew up a number of special terms and particular usages for otherwise common words which might be called the jargon of hop-picking.

The hoppers themselves used and understood this jargon without ever questioning it or stopping to wonder if anyone else knew what they were talking about. It was part of hopping, and ignorance of it marked the 'outsider' more certainly, perhaps, than almost anything else.

The following are just a few of the more common terms of the language of the hoppers. There are many more, applicable to other phases of the hop industry—planting, cultivating, training, stringing, stripping, drying—each developed its own vocabulary. Most of the special terms are as current today as ever they were, even though they are no longer used by the old hop-pickers at work in the gardens.

Alley

The space between two rows of hops was always called an alley. The bins were set into alternate alleys in order to give each bin a row of hops on each side of it.

Baskets

In some parts of Kent, particularly East Kent, hops were customarily picked into five-bushel baskets, rather than into bins as in South-West Kent and the Weald. The baskets had horizontal strings threaded across them which, when pulled, 'hovered up' the hops to give a lighter measure.

Bins

The most common container for the picked hops was the bin, which was also known outside Kent (in the West Midlands, for example) as a crib. Bins did not change very much in design and pattern, either from place to place or from year to year. Pictures of eighteenth-century hoppers show bins very much the same as those that were still being used in the 1960s. Basically, the bins were made of crossed poles or beams standing vertically at each end to form 'horned' legs. The paired legs were joined by horizontal beams about seven feet long, usually extending at each end to form carrying handles. From these beams hung a loose bag of coarse sacking which sagged to the ground and which could be divided in the middle—often were so divided—by a partition of similar sacking stitched all round to create two half-bins side by side. Pickers could take either a bin (usually for a whole family) or two half-bins (popular with one or two pickers). The crossed poles at each end of the bin were held in place by a bolt so that they could be closed scissor-fashion for moving the bin when empty and for storing between hop-picking seasons.

There is an early description of a bin in a book published in 1733 by The Dublin Society, called *On the Planting and Management of Hops*. That book described a bin as a frame of two poles or pieces of wood, nine or ten feet long and three or four inches thick, joined together about a foot and a half from each end by two other pieces three feet long, and supported by four legs three and a half feet high, leaving a space in the middle of this frame, six feet long, three feet wide and three and a half feet deep. In this space was fixed a coarse linen cloth or hop-bag, cut open on one side to lie hollow, hanging either by tenterhooks inside the frame, or stitched on the outside with wood skewers, to receive the hops as they were picked. Three men or women or four boys or girls were then regarded as the normal complement for each bin, standing on each side of the frame and picking hops from two poles at a time.

When hops were grown on poles (as they were in 1733), the first pole pulled for each bin would be stripped of its bine and laid across the top of the bin, its ends resting on the 'horns' formed by the crossed tops of the legs. Other poles, with the bines still wound round them, could then be angled over the bin's picking area, with one end resting on the ground and the other end laid on the top bin-pole. Pickers could thus stand against the hops to be picked and use both hands to snatch the hops off the bine and throw them, in one movement, into the bin.

When all the hops had been picked from one pole, then the bine would be stripped off the pole, and the pole was laid in the wake of the bin for eventual collection and re-use. Another pole would be all ready for the pickers to carry on picking from. It was the pole-pullers' or binmen's job to keep their companies of bins supplied with hop-heavy poles so that they wasted as little time as possible waiting for hops to pick.

After poles were replaced with string in the hop gardens, the top pole was no longer needed because the bine could be laid across the bin's horizontal frame beam, or across the lap of a picker perched on the beam.

Every picker had his or her favourite way of working. Some began at the top of the bine and worked towards the bottom, paying out the stripped bine across the bin. Some laid the bine across the bin and picked from the bottom, pulling the unpicked top of the bine towards them as they went. When pickers perched on the bin frame, one thigh hitched along its length leaving the rest of the same leg dangling, while the body was supported by the other leg held straight to the ground, the crossed legs of the bin behind the picker could be used as a backrest and the position made not nearly as uncomfortable as it sounds.

Bine

Hops grow on bines. Because of the obvious similarities between growing hops for beer and growing grapes for wine, many early writers confused the two by referring to hop vines. More modern writers have fallen into the same trap—George Orwell was one. They are wrong. Hops are the flowers of the bine.

Blowing On/Off

As hop-picking became more and more organized it became customary for pickers to take their places beside their bins each morning and await the 'blowing on' signal that meant they could start picking. On many farms, the old metal horn was used for the purpose, but a more usual signal, certainly in later times, was a whistle, blown by the farmer or measurer or someone in charge of the garden. The same sound signalled time to stop picking for the midday meal, time to start again after it, and time to finish for the day. Between-times there might be a 'blowing-off' for an interim measuring. At the end of the working day, the cry would be taken up by the pole-pullers and binmen in different parts of the garden relaying the instruction: "Pull no more poles!" (later up-dated to: "Pull no more bines!"). The object of this was twofold: to control the quantity of hops picked during any one picking period, and so regulate the supply of hops to the oast for drying, so that hops were not left lying about waiting to be dried and spoiling in the meantime; and also to make sure that all the pickers had equal opportunities to earn their money during the same picking periods. To be an experienced picker and able to pick more hops than anyone else was something in which to take pride, and by which to earn more from the season, but to work longer hours than anyone else, by starting earlier or finishing later, was strictly decried.

The blowing-off signal permitted a picker to finish picking the hops from a bine already pulled, but no more; and some pickers watched the clock—and the man with the whistle—very keenly when they knew it was coming up to time to stop. Then they would pull a bine, or even two bines if they thought they could get away with it (perhaps by making it look as though, in pulling one, the other came down with it, as sometimes happened accidentally anyway) and take whatever advantage there was from the extra picking. Some binmen were more strict than others with their pickers and, if they saw that sort of thing happening too often, would insist that the extra bines be left in the bin overnight. Since that meant that the hops would 'go off' and be both difficult to pick and give less than their fresh bulk to the bin's measure, this was unpopular. But a binman could always be sure of the backing of other pickers if he did enforce this ruling, because it meant that the advantage of the extra hops was lost, and the culprit would be discouraged from pulling that trick again.

Booking

When the hops were measured out of the bin, there would be among the measurer's little entourage, one man to hold open the poke into which each bushel was emptied from the measuring basket, one man to make sure empty pokes were ready for filling when one was full, and someone to do the booking. Each bin or half-bin was identified by a number and allocated a card with the bin-holder's name on it. At each measuring, the bin-holder would hand the card to the booker and when all the hops had been measured out of the bin, the measurer would write the number of bushels on the card. At the same time, the same number of bushels would be entered against the appropriate name and bin number in a master-book containing the names and bin numbers of all the farm's pickers. At the end of the picking, both farmer and picker thus had a record of the number of bushels picked by each bin-holder, and each was paid out according to the agreed tally, or amount of money per bushel.

Clothing

The hoppers did not use the word 'clothing' in any special sense, but most of them had special clothing which they set aside for the hopping. The black-staining resin from the hops quickly ruined any clothing that came into prolonged contact with it, so no-one who knew what to expect from a season's hopping wore anything but old clothes for work. Wellington boots were common footwear, even in dry weather; they were effective protection from dust and scratching bines as well as from rain and mud. Besides, even in the warmest hopping weather, the early mornings were often dew-drenched, and if there was an unexpected shower or more prolonged rain during the day, there was no opportunity to go home and change into more suitable footwear.

The rest of the clothing needed to be thick enough to keep

out the often sharp cold of late summer early mornings, yet discardable when the sun warmed the day into a heat that was often unrivalled throughout the rest of the year. So jumpers over shirts and blouses, and probably a coat or raincoat over that, were sensible. Men and women alike often wore aprons made of sacking specially for the hopping. Many women wore the popular floral print wrap-over overalls, sometimes with a sacking apron tied round the waist as well. It was these familiar overalls that did so much to give a hop garden its characteristically rag-bag colourfulness as the hops began to be thinned and the people spread over the bared earth.

Early mornings were often accompanied by very heavy dew or even frost and the hops would be wet and very cold so some hoppers wore gloves. The old hands despised such nonsense, declaring that they couldn't pick properly in gloves—which, certainly, restricted the nimbleness of the fingers. But some did allow themselves the comfort of a pair of mittens, often an old pair of the old man's woollen socks with the feet cut off, so that they covered the hands from knuckles to mid-forearm, leaving the fingers completely free. They did something to keep the hands warm, but they did rather more to prevent some of the soreness that resulted from the barbed caresses of the bines and the leaves.

Drifts

The picking of the hops involved a certain amount of organization, and the bins (or, in East Kent, the baskets) had to be marshalled into manageable units. So it was usual for bins to be set out through the hop garden, or part of the hop garden, in parallel ranks, anything from a couple of dozen or so to a hundred or more bins abreast. There might have been two, three or more of these drifts in one hop garden, each one a certain number of hills in front of the one behind it. Thus, each parallel line of bins worked through its drift in line abreast until it reached the picked area behind the line of bins ahead of it. After that, the bins would be moved, either being carried by the pickers helped by the binmen, or by wagon or trailer, to another part of the hop garden, or even to another hop garden on another part of the farm where they would be set out in a similar line-abreast formation in their respective drifts.

Heads
Hop bines often climb their strings (in the old days, their poles) and then, with nowhere else to go, simply thicken and tangle at the top. This thickening is known as the head, and bines with good heads were a mixed blessing to the old pickers. They meant a good picking from one bine, but they made that bine heavy to handle and it was not as easy to pluck the hops individually from a heavy bine as it was from a light one.

Heavy
Bines with good heads on them were said to be heavy. But the same term was also used of a particular style of measuring the hops out of the bins. If the measurer scooped out the hops into his bushel measuring basket, and used his hand to push the hops down to the mark, an inch or so below the basket's rim, which indicated the one-bushel level, he was said to be measuring heavy. Measurers were watched very carefully and if a picker thought the measuring was heavy, there would be a complaint about it. Some measurers were noted for their light measuring, but the only complaints about that came from the farmer. It was an indication of the measuring skill of a measurer if both farmer and picker judged him to be fair.

Hills
Hops grow on hills, or, rather in hills. Originally, as illustrated in old prints, they really were grown on hills, with the soil heaped up into two or three feet high mounds each supporting four poles (sometimes fewer) and the bines that grew on them. Today the hills are more like continuous ridges, but each group of four bines is still known as a hill.

Hop Dog
This was the name given to a particular implement used by pole-pullers. It was a large flat hook with teeth along the inside of the curve. The hook was fixed about eighteen inches from one end of a substantial pole from five to about seven or eight feet long. The hook was about nine or ten inches long, an inch broad and about three-quarters of an inch thick. In action, the toothed hop dog blade was struck into the hop-pole and then, with the lower end of the hop-dog pole resting on the hop hill,

the main length of the pole was used as a lever to loosen and prise the hop-pole loose from the ground.

The same name of hop dog was also commonly used, especially among the children, for a particular kind of furry caterpillar. Botanically it is the larva of dasychira pubibundal—and with a name like that it is not, perhaps, to be wondered at that it was much better known by the more simple and homely name of hop dog. Several different kinds of caterpillar are found among hops during the harvesting season, and the hop dog was one of the best known.

Housing

This was a term used, almost synonymous with heavy, for bines that had grown together at the top to form a thick canopy of leaf and hops. Whereas hops that were heavy were, by and large, welcomed, in spite of their disadvantages for the picker, hops that were housey were generally disliked because a lot of the hops tended to be lost and wasted when the bines were pulled and the entwined tops were forced apart.

Hovering Up

When it was time for the hops to be measured out of the bins, the bin-holders were expected to help the measurer by hovering up the hops to one end of the bin so that instead of having to dip the measuring basket into a flat surface of hops, the basket's mouth could be held against an inclined surface and the hops brushed lightly into the basket. In this way, less damage was done to the hops during measuring, and the measurer's back did not have to bend quite so far as the level of hops fell. As the hops were taken out, the bin-holder would bend over the bin, usually from the side opposite that from which the measurer was working, thrust his or her forearms down into the hops in an embracing motion and hover up the hops ready for the next plunge of the measurer's basket. In this way, the inevitable compressing effect of the basket was countered and the hops allowed to fill the basket as lightly as possible.

In the East Kent five-bushel picking basket regions, the baskets themselves were strung with hovering strings which could be pulled to loosen the hops in the basket and, just as the arms of the bin-holders did in West Kent and the Weald, prevent the compacting that made the measures heavy.

Light

Hops were said to be light if they grew more or less evenly on the bine without crowding on the overhead wires. Often a light crop meant that the individual hops were larger, so that the measure would not be all that much different from a heavy cropping bine. Light bines were much easier to pick and hopping progressed much faster when the hops were light than when they were heavy.

Light measure was achieved by a measurer who habitually scooped hops out of a bin and then used his hand and forearm to brush surplus hops out of the top of his basket to expose the bushel mark that ringed the basket just below its rim.

Measuring

Measuring was a skilled job, calling for a degree of physical fitness, a great sense of justice and fair play, and a certain amount of firmness, too. There were always pickers ready to challenge the measure. Bins were measured out two, three or sometimes more times a day, usually according to the needs of the oast house where the drying was going on all the time. The measurer would arrive at each bin with a little entourage of attendants, binmen and booker. The measurer had only his measuring basket, designed to hold just over a bushel if filled to the top, but made with a dark band an inch or so from the top to show where the hops should come to in order to measure just one bushel. Because the pickers were paid according to the number of bushels they picked, the measurer's interpretation of what was a bushel was critical. Before he arrived at the bin, the bin-holder would hover up the hops, cleaning them of leaves and bunches and bits of bine in doing so. An experienced picker could make an appreciable difference to the measure, and therefore to the value of the tally, by skilful hovering up. When the hops were ready (and that had to be before the measurer arrived; he was not a man to be kept waiting at the binside for any reason) the measurer took his stand beside the bin. Beside and slightly behind him would stand one of the binmen, holding a ten-bushel hop-poke open at the ready to receive the hops from the measurer's basket. Near him, a second binman would stand ready to take the filled poke and hand out an empty one to replace it, so that the measurer's rhythm did not have to be interrupted. The second binman would have a supply of hop-string, conveniently cut into suitable lengths ready, stuck through his belt so that, having handed over the empty poke he would take the full one, draw the neck together and tie it firmly with one of the lengths of string and leave it where it could be collected by the wagoner for conveying to the oast.

The booker, meantime, collected the bin-holder's tally card, bearing the number and holder's name of the bin. On the card would be entered the date and the number of the day's measuring (first of the day, second, third, and so on). As the measurer took the hops out, he would sing out the number of bushels and when he finished with the bin, the total number would be entered on the bin-holder's card, and also into the big book that was the farm register. As the measurer moved on to the next bin, the booker would hand back the card to the bin-holder and move after him, ready to do the same for the next bin.

Oast Hairs

Not, strictly, a term that was much used by the hop-pickers themselves, many of whom never had any reason to know in any detail what went on anywhere once the hops left their bins, via the measurer's basket. But oast hairs were the horse-hair blankets that were spread across the laths of the drying floor in the oast kilns. The green hops coming in from the gardens were spread on the horse hair, about six or eight inches deep at a time. The fire below would send the hot air filtering up through the laths and the blankets and through the hops themselves, drying them evenly and gently until it was time for them to be turned and, after that, to be unloaded out of the drying kiln on to the barn floor ready to be pressed into the pockets, in which the hops went to the markets.

Paying Off

The practice for paying off the hoppers varied from farm to farm in detail, but broadly it was always the same. Sometimes, the paying-off office was set up in the oast house. Sometimes a room in the farmhouse would be used for the purpose. If the weather was good enough, it might be an open-air occasion. However it was done, it was always an occasion—more than just the end of hopping. The farmer would sit behind his table, often with a cash-box at his side, and the bookers' farm register in front of him. Usually he had an assistant to check his calculations and help have the money ready for each picker. Sometimes, there would be a local policeman, doing his best to look as though he had just dropped in to see the annual ritual. The hoppers would line up ready to file past the table, or hang about in groups waiting for their names to be called. Then they would present their bin-holders' cards, which would be checked against the farm register. If the two told the same tale—as they should do—then the total number of bushels picked during the season would be announced and the appropriate earnings, at so much a bushel (or so many bushels for so much), would be handed over. There was usually an exchange of pleasantries, thanks from the farmer and a hope that he would see the picker again next year, and contractual obligations between the two were at an end.

In fact, the foreigners often stayed on the farm for another day or two, making ready for the return home and the night after paying-off was, in some cases, the traditional occasion for the hoppers' feast.

POCKETS

Hop pockets are long sacks into which the dried hops are pressed ready for marketing. A filled pocket contains about 112 pounds of hops and has to be moved about on a little trolley when it cannot be manœuvred by hoist and tackle.

POKES

These were much coarser sacks, usually brightly coloured (orange, generally, but ranging from yellow to red and occasionally greeny-yellow) into which hops were measured from the bins. Each poke held ten bushels. When a measuring was due, a load of hop pokes would be brought to the hop garden from the oast house and dropped in small bundles in the alleys behind the bins, where they could be gathered up ready for filling by the binman whose job it was to keep the measurer supplied with empty pokes to fill. Once filled, the pokes were again left behind the bins, and the wagoner would drive along again, once measuring was over, and the full pokes would be loaded up to be taken to the oast. The pokes, with their ten bushels of green hops each, were quite light, and even the old horse-drawn wagons would take an enormous load of pokes, piled high. It was always something to be remembered by children to be tossed up on to the top of the load for a ride, although such a privilege was always attended by warnings about sitting still so as not to crush the hops.

Empty pokes were used for a variety of purposes in the hop gardens. They were ideal for sitting on during the midday meal break (if any happened to be lying about at that time). They were used to mend the bins, too, and if it rained, the measurer's crew used them to make shoulder-covering hoods by pushing one bottom corner into the other.

Pole-puller's Pole

Often known as a hop-dog, but strictly a different implement. It was a pole ten or twelve feet long, quite thin (perhaps an inch in diameter) with a small curved blade, like that of a small brishing hook, fixed to the top. The blade had a spur-like spike sticking out of its heel, sometimes flat, sometimes round. The blade was used to cut bines that had entangled a. the tops of the poles or, later, the stringing wires, so that the bine could be brought down without too heavy a shower of wasted hops. The spur was used to hook free any part of the bine that was left aloft. It took a certain amount of practice to manipulate one of these poles successfully, and a certain amount of strength of arm, too.

An unintended and probably quite unauthorized use for the pole was as a means of keeping some of the younger children in order. There was something specially awesome about a pole-puller, who could make his presence felt and his authority recognized from twelve feet away.

Poles and Stringing

"If I could discover an everlasting hop-pole and one, too, that would grow faster even than the ash, would not these Kentish hop-planters put me in the Kalendar along with their famous Saint Thomas of Canterbury? We shall see this, one of these days." Such a bold prophecy, made by William Cobbett during one of his Rural Rides (1821–32) deserved to be fulfilled. In a way, it was.

Not exactly, though. There never did come into the hop gardens an everlasting hop pole. What did come, though, was

Hoppers at their bins near Yalding in the 1930s

The tradition of picking into baskets in East Kent was carried on until the end of hand-picking

Conveyor-belt hopping has replaced the older methods during the second half of the twentieth century.

The modern equivalent of a bin-man cuts down the bines from a tractor-mounted 'crow's nest'

string and wire. The hop poles—mainly larch, chestnut and ash—lasted for nearly 300 years. They were used in the first English hop gardens and they were still being used in many gardens well into the twentieth century. Different varieties of hops needed poles of different lengths. At first, the poles were set three or four to a hill, and the hop plants in the hills were trained up the poles. There were different arrangements of poles, according to the way in which the hills were planted. Rows could be six, seven or eight feet apart, and whichever was the case decided how the poles should be set. A six-foot plant would have the hills six feet apart, 1,210 hills to the acre and three poles to every hill. A seven-foot plant would have four poles to the hill, 889 hills to the acre; and an eight-foot plant put the hills four feet apart, two poles per hill, 1,361 hills to the acre.

Later, other arrangements of poles were devised. A Maidstone farmer was credited with one which became popular: it used two poles, each eight feet long, permanently in position each side of each hill. A socket was fixed to each of the poles, four feet from the ground, and wire was bound to the tops of the poles to form a fork. Then, 12-foot long poles, their butts resting in the sockets, their tops in the forks of the opposite uprights, linked the two. For picking, the 12-foot diagonal poles could be lifted out of the sockets and forks and laid across the bins. The hops seemed to grow particularly well using this method and it made unnecessary the pulling of poles, which often broke them so that they had to be replaced the following year. But it did make tying and training more difficult, and therefore more expensive.

Another farmer early on tried training his hops on an arrangement of vertical and horizontal wires supported on stout poles at the end of each row, which was simple, and economical, but in the end, it was the American innovation of overhead wires and vertical string that won the day and left us with the various cats' cradle patterns that are the familiar appearance of all hop gardens today.

Sets

A set was a group of bins, the number varying, in the charge of one pole-puller or binman. It was also the area of ground at which the bins were working at any one time. It was, therefore, usual to refer to the bins working through a drift in sets.

Shoe-money or Foot-money

This was an old hop garden custom, apparently peculiar to Kent, dating back to the eighteenth century at least. It required all visitors to the hop garden to pay a token toll for the privilege of having their shoes wiped with a bundle of hops. As soon as a visitor was sighted, pickers would hurry to exact the toll which, however, was not normally a personal 'perk' but a contribution towards the cost of the traditional end-of-hopping feast.

Snatching and Stripping

Two of the techniques for picking hops from the bines. Experienced pickers laid the bines across their laps, lifted a branch which they held clear of the rest of the bine in one hand while they snatched off each hop individually with the thumb and second finger of the other hand. Some of them could do this at incredible speed. But sometimes a branch would be suitably shaped and sufficiently free of leaves for the hops to be stripped off in one action, by closing the thumb and forefinger into a circle at the base of the branch and simply drawing the hand along the branch, from bottom to top, so that the hops were stripped off. This, of course, was quicker and experienced hoppers could even do it without plucking the larger leaves off the branch at the same time. In practice, picking the hops involved a combination of the two techniques.

The ease with which different varieties of hops could be snatched off the bine was, in fact, made the subject of a scientific study in 1951, when it was found, for example, that the Canterbury Golding needed a pull of 56 grammes to separate the hop from the bine, whereas an Eastwell Golding needed a pull of 332 grammes, and a variety called Bullion needed 494 grammes. The same study showed that the ease of picking depended more on the size of the hops and the thickness of the clusters on the bines, whether they were wet or dry, how leafy they were—and a number of other elements.

Tallies

Tally sticks were the earliest means of recording the number of bushels of hops picked by a hop-picker, and they were in use from the sixteenth century right up to the twentieth. Basically they were one piece of flat wood, sometimes rounded at the end.

Two or three inches from one end the wood was sawn half through its thickness and then split down its length, so that the two pieces could be laid together again into an exact fit. The two pieces would each be branded with the same number, and the shortest piece would be held by the picker, while the remainder of the original piece of wood was threaded on a cord and given, with all the others (one for each bin or half-bin), into the safekeeping of the tally-man. At measuring time, the tally-man would separate the appropriate tally stick from the rest on his cord as he arrived at a bin, collect the corresponding piece from the picker at the bin and, placing the two sticks together, he would mark both of them across one of their sides.

These tally sticks were, in fact, basically similar to those used by the Exchequer to keep the nation's accounts until the end of the first quarter of the nineteenth century. In the hop gardens, the tally-man might also be the measurer, but usually he accompanied the measurer, just as, later, the booker did. The mark he made across the thickness of the two halves put together were made with a triangular file.

After the introduction of hop tokens, tally stick and tokens were used together, so that the picker would receive a token for one to four bushels, and a notch in his tally stick for every fifth bushel. For a twenty-four bushel measuring, therefore, he would have his tally stick marked with four notches before it was handed back to him, together with four tokens, which were a sort of small change. Gradually, the tokens, which were much more convenient for everyone, took over from the tally sticks altogether before they, too, were taken over by cards and a register.

Tokens

Hop tokens were a feature of hopping in Kent and Sussex for at least two hundred years, during which time they were both a means of reckoning how many bushels of hops were picked by a picker, and also a form of currency, not only in the hop gardens but, certainly in some cases anyway, in the neighbouring villages, too.

One of the first—perhaps the first—to take a collector's interest in hop tokens was the Reverend R. W. H. Acworth who, between the wars, went around the farms of Kent and Sussex seeking out forgotten hoards of tokens and patiently piecing them together into sets. There are a number of these sets in Kent now. The Kent Numismatic Society has one, and there is an interesting collection of them in Maidstone Museum.

The tokens vary a great deal in artistic merit. Some are very crude—mere discs of light alloy stamped with a figure denoting the number of bushels the piece represented. But some are really rather beautiful pieces, attractively decorated and lettered, and obviously the work of craftsmen. We know of one such craftsman: a Mr Comport of Northiam, who made tokens for farmers in his area, and later machine-made pieces were being supplied by S. A. Daniell of St Paul's Square, Birmingham, who made sure his name and address figured prominently on his customers' tokens. Often, the initials of the hop grower issuing the tokens were incorporated in whatever markings there were, and occasionally the full name of the farmer was cast into the metal.

The sets tend to run in tokens of 1, 6, 12, 30 and 60 (with an occasional 120) value. Smaller value tokens would have been exchanged for larger ones as the hopping went along, and whatever token value a picker was able to hand to the farmer at the end of the picking, would represent the number of bushels for which payment was due at the agreed tally.

The earliest tokens were cast in pure lead. They were very simply made, probably poured from a ladle of molten lead into a disc-shaped mould cut or perhaps burned into a piece of wood. They would have been made by the farmer on the farm. The usual pattern was for the small denomination tokens—the one-bushel pieces—to be very small indeed (about the size of a penny piece today; sometimes even smaller) and the tokens increased in size according to the number of bushels represented. That number would be punched into the metal of the token after it was cast. Sometimes, identification was made easier by varying the shape of the smaller tokens, so that some of them are, for instance, square, instead of round like the larger ones. No-one knows when the practice began so there is no way of telling quite how old the oldest of these lead tokens might be. Gradually, though, and by no means universally, the quality of the best examples of hop tokens improved. It became customary to punch the growers' initials on one side, while the denominational figure was punched on the other side. Sometimes the figure was accompanied by a letter B (for bushels);

sometimes by a D (for dozen bushels), but the absence of either of these does not always mean that the figure is intended to be one or the other.

By the beginning of the nineteenth century, pure lead was being replaced by lead alloy, which was harder-wearing and lighter. Many of these were produced for the farmers by manufacturers and these are among the most attractive examples of this curious branch of the coin-makers' craft. By the mid-nineteenth century another change of fashion introduced thin discs of iron washed with tin, which generally looked cheap and more or less nasty, but probably served the purpose just as well.

It is impossible to date any of the tokens accurately by the materials in which they are made, or the degree of art and craftsmanship that went into their manufacture. Throughout the life of hop tokens individual tokens were made of lead, lead alloy, tinned iron, bronze, brass, bell-metal, copper, pewter, zinc and even cardboard and paper. There is no doubt that some of the best pieces were used by farmers as a kind of status symbol.

Luckily, though, some farmers actually dated their tokens, so we know for certain that they were in use, at least in some parts of Kent, in 1767. Some of the tokens had holes in them, and on some farms it was usual to issue pickers with wooden meat skewers so that when the measure was taken the hops were paid for with a holed token which could be slipped over the wooden skewer for safe keeping.

Probably, most of the tokens outside the known collections have long since been thrown away or perhaps recycled in some way. Some old hoards may well have been turned out of barns or oasts during either of the two World Wars to be melted down for munitions. Certainly, when they went out of general use around the time of the First World War there must have been quite large quantities of them in boxes or bags on most of the Wealden farms. There may well be several caches even to this day in old farmhouse lofts, or other hidey-holes.

The use of the tokens was widespread. They have come to light on farms in Worcestershire and Herefordshire, as well as in Kent and Sussex. Of course, although they are called hop tokens and were undoubtedly used for hop-picking, they may well have been used at other times of the year, too. The use of tokens was common on farms where hops were not a specially important crop as well, and it is not unlikely that on some farms, the tokens were designed to be used as general purpose pieces throughout the year. They would, after all, have been redeemed by the farmer at the end of the harvesting of each crop, in exchange for coin of the realm, and so would have been available for use—although in smaller quantities—for fruit-pickers, pea-pickers, and, in fact, for any of the many other piece-work harvesting jobs. The original purpose of the tokens has not altogether died out even today. On some farms, piece-work earnings are still reckoned by tokens handed to the workers and redeemed at the end of the season or the end of the week for a cash equivalent. Nowadays, the pieces are often made of plastic and (just as the original pure lead tokens were) are of different shapes and sizes and (oh, yes, things do change!) different colours too for the different values.

TWELVE

The Volunteers

Father Wilson, as he passes,
He smiles at the lads and lasses.
He is so good and kind.
At the Rose and Crown, at Hopping Time
With his ABC cakes and his cup of tea,
He fills the hoppers' hearts with glee,
By Jove! it ain't half fine
At the Rose and Crown, at Hopping Time.

Part of "The Hopping Song", composed by one of the patrons of the Rose and Crown, The Little Hoppers' Hospital at Five Oak Green

Sir Kingsley Wood, Parliamentary Secretary to the Ministry of Health, after a visit to South-West Kent hop gardens in 1927, commented upon the improvements that had taken place in hoppers' conditions. "They owe a great deal to the splendid voluntary organizations which are in existence," he said. And indeed they did. It was the voluntary welfare workers who drew attention to the plight of the hoppers and, more important still, who did what they could to relieve that plight, even as they battered away at the sluggish social conscience of the day to try to get a wider recognition of the need for improvements.

The arrival of thousands of men, women and children, in a comparatively small area of Kent for perhaps a month of each year, set the local authorities in the receiving areas no small problem. At the beginning of the nineteenth century, almost the only local authority concern for the hoppers was that which was thrust upon those responsible for Poor Law administration, and they tended to be more concerned with finding excuses for not accepting any responsibility than for doing anything more positive.

Undeniably, conditions in the Kent hop gardens throughout

most of the nineteenth century were bad. Accommodation was poor and often wholly inadequate, so that there was unbelievable overcrowding. No arrangements were made to collect the hoppers' rubbish from the camp sites, so it lay about where it fell and rotted there. Lavatories were incredibly primitive, even when there were any; water had to be carried from a well or a pump a mile or more away in the farmyard in many cases, or else taken from a more convenient pond or ditch of very suspect cleanliness.

Conditions encouraged illness of all kinds among hoppers, yet, when they fell ill, the only doctor who was prepared to treat them might be miles away in the nearest town, and the only way the patient could get to him would be by walking.

It was left to philanthropic volunteers to make the first effective moves towards improving the hoppers' lot. In some cases, the moves began with local people, usually people moving into the hop garden area who saw things without the blinkers of life-long custom. Sometimes, curiosity was aroused in London about where these thousands of very poor people went every summer and what they went to.

By the middle of the second half of the century, organizations were being set up specifically to help the hoppers during their annual stay in Kent but the help they were able to give was, at first, very limited. Many of those first voluntary workers were churchmen whose first concern was for the souls of the people they wanted to help but they came to recognize that the needs of the body were as pressing as those of the soul, and there began a period of 'pound of tea' religion, in which the bodily appetites of the hoppers were used as a means of access to their spiritual needs. After that, it was inevitable that successive workers came to understand that the great need was not just for tea, buns and a Bible text, but radical reform of the whole way of life to which—and even from which—the Londoners came.

It is easy enough now to blame the farmers of the day for not rushing to bring about the improvements that began to be urged upon them but, as the reformers themselves quickly discovered, the hoppers, although they were well used to grumbling and squabbling among themselves, were quite unused to organized protest, and so did nothing at all to support their reformist spokesmen. Instead, many of them regarded the 'do-gooders' with a great deal of suspicion, as outsiders poking their noses in where it couldn't possibly do any good, and might do a certain amount of harm by way of reaction among the farmers.

The Londoners of the period had learned, through generations, that in almost every way throughout life the cards were stacked against them. Many of them were illiterate, or almost so, and virtually wholly uneducated. They were hedged about on all sides by an uninterested and even antagonistic Authority over which they had no hope of prevailing. Their instincts were all directed towards closing ranks, not making waves which experience had taught them would only be turned back on them, and generally giving as little trouble as possible. It didn't matter very much how bad things were; they didn't expect anything different. They were 'the poor', and could not hope for anything better than whatever life meted out to them.

So the reformers had not only to try to persuade the farmers to offer their hoppers something better than they were used to offering them; they had also to try to get the hoppers to want something better.

It must have been very uphill work, and it is no wonder that some of the reports of those early reformers read like accounts of Christian martyrdom, with missionaries telling of how they were spat upon and reviled as they went among the hoppers in their camps and in the hop gardens, and at the same time treated with barely polite tolerance by many of the farmers.

Their representations to various authorities, national and local, were received with the same sort of polite off-handedness, and for years the missionaries were praised publicly for their selfless concern for others, but privately very largely ignored. There was even a feeling in some quarters that to behave any differently would rob these well-meaning people of their opportunities for doing good, so that it was, therefore, the right and proper thing to praise them with words and do nothing more.

Still, things did get done. The missionaries recognized the need to comfort bodies first and save souls afterwards, and they began to provide medical services for the sick, food for the hungry, and advice about domestic problems to people who regarded their worst problems as just an ordinary part of everyday life.

The overcrowding in the hopper huts was one of the chief

targets of the early reformers, who saw this feature of hopping down in Kent as a frequent beginning to a life of shame and delinquency for many a young Londoner. As late as 1892, a Salvation Army report commented: "Many a hitherto pure girl goes back to the city contaminated, to drift ultimately to the streets."

The drunken habits of the hoppers appalled many of the missionaries, and other people, too. In September 1881, a letter to *The Times* wondered "how it was possible that, in a Christian country, a 'considerable County town' could be thronged with drunken men and women, and the side streets and alleys blocked with human beings prostrate with drink".

The 'considerable County town' was, in fact, Maidstone.

One of the very earliest campaigners for improvements was the Reverend J. J. Kendon. He saw for himself what things were like in the Kentish hop gardens when he first arrived at Goudhurst as vicar there, and he determined that something must be done for these poor people. To be sure, his first concern seems to have been, understandably perhaps, with their spiritual rather than their physical condition. He found them to be "destitute in spiritual things" and in the 1860s he started a mission near Goudhurst, at Curtisden Green.

Kendon's work was chronicled by the Reverend R. Shindler, particularly in his book *Mission Work Among the Hop-pickers in the Weald of Kent* which was published in the last decade of the nineteenth century. In his book, Shindler commented:

> "The Gospel is, distinctly and emphatically, the weapons of the warfare of the Agents of the Hop-pickers' Mission; it is the message they bear to the sin-sick, the sorrowful, the guilty and the perishing among whom they labour. . . ."

Kendon himself soon realized that his efforts to bring the Gospel to the Godless in the hop gardens were likely to be much more successful if he could first, or at any rate simultaneously, demonstrate a genuine recognition of and concern about their human condition and needs for decent food, clean water, proper accommodation, and so on. Soon, he and his supporters were campaigning among the farmers and any influential friends wherever they found them to bring about the improvements they advocated.

In one of his reports, Kendon wrote:

"We go into the hop-gardens, talking to the people from bin to bin, inviting them to tea, giving tickets to as many as can come. We find the people very susceptible to the kindness shown to them in this way. . . ."

And he went on to say:

"Whatever may be lacking about these poor creatures, you seldom find them destitute of a good appetite. The amount of provisions which disappears is remarkable."

The hoppers who accepted these invitations to tea could usually be prevailed upon to stay for the prayer meeting that followed it, during which there would be prayers, a hymn or two and a short address. Most of the Mission's guests were children, but many of them were accompanied by one or more adult, too, and so the purpose of the exercise was accomplished. As time went by, these mission prayer meetings became a regular and even popular feature of the hopping season throughout the Kent hop gardens.

The Reverend J. J. Kendon gathered a little band of helpers around him and more missions were begun in other parts of the Weald. He founded one at Horsmonden in 1868, where he acquired the use of a small hut and the services of a nurse to give medical aid to any who came to ask for it. Soon there was a Weald of Kent Hop Pickers' Mission medical hut in most of the hop-growing area around Goudhurst, where the work began, all supported by voluntary contributions of money and effort, each one staffed by a London missionary for the season.

When he died, his son, Mr Samuel Kendon, also of Goudhurst, carried on the work, until, in 1921, it was taken over by the Reverend F. J. Feltham at Horsmonden and from that time the Horsmonden Mission, although still attached to the Weald of Kent Mission, operated independently. As time went by, the original work begun by the Kendons, father and son, was taken over by the British Red Cross Society.

In the meantime, the Reverend F. J. Feltham carried on the good work from his Horsmonden base. It was he who asked Southern Railway to change the times of the hopper specials so that they no longer left London during the night and dumped the hoppers at Kent villages before daybreak. As a result of this plea, the times of the trains were changed and the hoppers did not arrive at their destinations before 6 am. That still meant that many of them had to wait about until 7 am or even later, and the Mission began to arrange for the early trains to be met at the stations with tea and other refreshments for the disembarking passengers.

The railway authorities gave permission for a canteen to be set up in the station yard at Horsmonden, and the voluntary workers were allowed to go on to the platform with a trolley. There was also a canteen on the village green, and many hoppers came there for their midday meal during the hopping. When it was time for the hoppers to go back to London, the Mission trolley was again on the platform and the canteen in the station yard, waiting to serve them and send them off at least with a comforting warm drink inside them.

In 1924, Feltham put up a dispensary hut on the Brenchley Road at Horsmonden, to serve some thirty-six neighbouring farms, and when he died in 1931, his work was continued by Mr A. Hill, who had been the Superintendent of the Horsmonden Mission for ten years and had had, altogether, thirty-three seasons' experience of the work.

Soon after that, the nature of the work of the missions began to change. By 1954 there was no longer any need for the canteen work. Very few pickers were arriving by train any more; most of them came by lorry or car, complete with vacuum flasks and picnic equipment.

It was largely as a result of the work of the Reverend J. J. Kendon in and around Goudhurst that Kent hop-growers banded together to form The Society for the Employment and Improved Lodging of Hop-pickers. The Society was formed in 1866, and members drew up recommendations for improvements which formed the basis of model by-laws suggested by an Act of Parliament in 1874 which gave local authorities powers to make by-laws covering, primarily, hop-pickers' accommodation.

Unfortunately, although they had the powers, not all local authorities had the will to make those by-laws—at least, not in the recommended form. For one reason or another, some of them decided to make by-laws of their own, different from the models provided in the Act, which in practice left loopholes that meant the hoppers were no—or very little—better off than they had been before. Certainly, the Act did nothing at all to remove the need for the services provided by the voluntary organizations, although it did serve to give official recognition to the need for improvements and helped to publicize that need. The number of voluntary organizations that began to take an active part in hop-garden welfare work of one kind or another increased rapidly after 1875.

The Society for the Employment and Improved Lodging of Hop-pickers, however, did not fade out of existence just because the recommendations it had made were not adopted by the local councils. Many of its members were very influential land-owners who fully recognized their moral responsibilities towards the pickers and it was the Society that first introduced a system of hiring hop-pickers through trusted agents living in the pickers' home towns or districts of London, and that alone did much to regularize the arrivals of hop-pickers into Kent and so to prevent at least some of the worst of the former over-crowding.

The Church of England Missionary Association for Hop-pickers was formed at Maidstone in 1877 to co-ordinate the work of the various missions which by then were beginning to make their presence felt throughout the Weald and South-West Kent generally.

In 1889 the Church of England Temperance Society turned its attention and the very considerable energies of its supporters to work in the hop gardens. The first emissaries of the Society were mostly ladies who went out into the hop gardens dressed, very often, as though for afternoon tea on the vicarage lawn, in laundered blouses and ankle-length skirts and straw boaters. They took with them little three-wheeled barrows loaded with jugs and mugs of hot tea and coffee, buns and cakes, and very soon the ladies had become familiar figures in the hop gardens of both East and West Kent.

In 1901, the CETS also began to set up stores in tents, and to organize proper supplies for the hoppers who, until then, had had to rely on local shops, often some distance from the hop gardens and, anyway, closed by the time they got back to their huts for the night, or upon itinerant traders who, frankly, fleeced their customers by passing off food of very doubtful quality indeed.

The CETS, too, had its own medical services and the first CETS travelling dispensary was a box on the back of a bicycle pedalled by a lady doctor who went from hop garden to hop garden offering medicines and ointments for all sorts of minor ailments, from upset stomachs to bites and scratches. By 1958 the CETS was running twenty-five dispensaries and covering an area of 200 square miles of hopping country with its medical service.

The CETS ladies actually began their hop-picking mission work in the Canterbury area, around Harbledown and Selling, taking their hot tea and refreshments to the hop-pickers in the gardens every morning and afternoon. They trudged many miles every day over farmland and lanes pushing or pulling their little coffee-carts or carrying great steaming jugs of hot drinks. Eventually, they set up static refreshment centres where they could brew up and dispense their goodies more easily. They opened grocery stores in several of the larger camps in the Wealden areas, from Paddock Wood to Yalding, and these stores also served as first-aid posts from which the ladies could offer social and evangelistic work, as well. For these workers were, first and foremost, evangelists, spreading the Gospel in a practical way.

Most of the ladies lived locally, but they were joined by CETS colleagues from London, too. One of these was Dr Elizabeth Taylor-Jones who for many years spent her annual summer holiday offering medical and spiritual help among 'her Cockneys' while they were hopping in Kent. She became an almost legendary figure, known to thousands of hoppers as she toured the hop gardens in her car, arriving with a strident invitation to "Bring out your dead!" It was the sort of humour the Londoners understood and enjoyed, and they responded in like vein, even on one occasion bringing a sheeted figure on a stretcher to her.

Another legendary figure of Kentish hop lore is Father Richard Wilson of St Augustine's, Stepney. Stepney, of course, was one of the South London areas from which many of the hoppers came into Kent each year and in 1896 Father Wilson decided that he wanted to see for himself just what it was his parishioners migrated to each year. He had, of course, talked to some of them about the annual exodus but this was the time before hop-picking had become the twentieth-century holiday with pay and the only information he got from the hoppers he spoke to in London was: "Hopping is hell!"

Anxious to find out for himself just what hopping was like,

Father Wilson persuaded one family to allow him to go with them. And so, in 1896, he joined the throng of men, women and children with their old prams and boxes on wheels all making their way, in the very small hours of a late August morning, to London Bridge Station.

The train—a hoppers' special—took them to Paddock Wood, where everyone got off and, still with their bundles and their prams and push-carts, began the long trek to the farms of the neighbourhood. The family Father Wilson was with were bound for a farm near Five Oak Green and when they arrived there, the priest settled in with the unmarried men in a barn. During the season, he lived just as the hoppers themselves did, picking at the bin all day, stumbling along the lanes and cart-tracks back to the barn at night.

He watched, and he listened, and he learned a great deal about hopping down in Kent. One of the things he observed was a dead man's body lying covered with sacks in a ditch. Another was that the only drinking water for the camp had to come from a dirty pond.

Most important of all for the future, though, was his chance meeting one day with a woman carrying a brown paper parcel. The woman told him the parcel contained the body of her baby, which she had carried five miles to a doctor, only to be told when she got there that the child was dead.

Father Wilson was shocked. "This must never happen again", he said and he made the exclamation a vow. The next year he came back to the Kent hop gardens and to Five Oak Green, only this time he had with him three nurses. The four of them hired a cottage for two and sixpence a week, and there they set up the first Little Hoppers' Hospital, primarily intended to look after the little hoppers—the children who came with their parents to the hop gardens, and who fell sick.

There was plenty of work that first year, for there was a smallpox epidemic, and Father Wilson and his nurses were kept fully occupied throughout the hopping period. Of course, they could not do all that they would have liked to do, or all that they saw needed to be done. Nevertheless, when the hopping ended, it was clear that they had met a desperate need, and Father Wilson determined that the work at the Little Hoppers' Hospital should go on.

The next year he was back, this time in a larger rented

cottage which, however, was still not large enough to cope with the demand for his services. In 1910 he was able to move the hospital into an old, disused public house, the Rose and Crown. The wooden skittles alley of the old pub was the first children's ward there, replaced in 1927 by a brick building, the Richard Wilson Memorial, still identified today as the Rose and Crown but known wherever hoppers live as the Little Hoppers' Hospital.

Father Wilson limited his area of work to a very small area in the Five Oak Green and Capel area, recognizing the limitations of his own resources and the enormity of the problem he was trying to tackle, but his influence extended far beyond the hospital's immediate catchment area and his work as well as his reforming zeal did a great deal to help bring about the changing conditions of the period. He was trusted and admired by the farmers and by the local authorities, as well as by the hop-pickers, and so he was able to help all of them to work together to bring about the much-needed improvements. It was he who first introduced the London British Red Cross Society to Tudeley and Five Oak Green area and he was, almost certainly, the main inspiration for the several College missions that started their own work in the early years of the twentieth century.

Both the Salvation Army and the Church Army had their workers in the hop gardens. The Salvation Army was publicizing conditions there in 1892, condemning the often inadequate weather-proofing and the almost unbelievable crowding together of men, women and children. "Much evil comes of it," said one writer in the S.A. magazine, the *Deliverer*, and he went on to detail the drinking, the foul language, the quarrels and the frequent ill-treatment of women and children that was commonplace among the hoppers.

"Indecent habits are common and open, and many a hitherto pure girl goes back to the city contaminated, to drift ultimately to the streets."

As is the habit of the Salvation Army, propaganda was backed up by practical work among the hoppers, and one of the features of many a hoppers' encampment at the turn of the century was the Salvation Army lasses looking after hoppers' children in their crèches, while their parents got on with the job of picking the hops.

The Church Army sent mobile units into the hop gardens, based with specially equipped vans, but some of the workers had bicycles, too, and toured from hop garden to hop garden stopping to chat, offering advice and a friendly word and generally doing whatever they could to make hopping a happier time than it had been in the near past. In the evenings, they offered first-aid treatment—usually of the cure-all pink medicine and bandage kind—and held services for children and adults, concerts and—a Church Army speciality—optical lantern meetings. These were early film shows, with slides instead of films, and very popular they were, too.

Many of the services and some of the meetings were held out of doors, around camp fires. At least one Church Army group, in the Cranbrook area in the early 1920s, organized a concert party which toured the camp sites with a piano on a lorry.

The University Colleges also sent doctors, medical orderlies and students to found and maintain, year after year, missions on particular farms and in particular hop-growing areas. One of many such missions was established by the Reverend John Campbell, who was Chaplain of Hertford College, Oxford, from 1909 to 1924. The adoption of a hop farm by the College was first suggested at a meeting in November 1910, by an undergraduate, D. L. Russell. A committee was set up by that meeting to consider ways of alleviating the social problems that attended the annual hop-picking, and the adoption of a farm was agreed upon. The first camp was held in the following September at Crowhurst Farm, Peckham Bush, Paddock Wood, with the Reverend John Campbell in charge. There were about a dozen members of the College at the camp. They paid all their own expenses and offered medical and other help wherever they could.

The experiment was successful enough to be repeated in 1920 and after that it became a regular annual feature at Crowhurst Farm up to and including 1924. Then Mr Campbell left to become Principal of Trinity College, Kandy, in Ceylon, and the camp mission ended.

Mr Campbell and his band of helpers were apparently able to make themselves welcomed by the hoppers who, after the Hertford College mission ended, felt that they had been left without a friend they had grown to expect to be there each year. So, in 1926, the camp was started again, this time by

Miles Sargent, who chronicled his experiences there in his book *St Francis of the Hopfields*, which was published in 1933.

Father Sargent's mission staff was made up of Oxford undergraduates and medical students from London hospitals. Starting with a staff of half a dozen, the number was increased year by year to more than twice that number. They worked a fourteen-hour day—often more—and lived pretty roughly, deliberately choosing to live in no greater comfort than the hoppers themselves did.

They built themselves a hut of corrugated iron to live in, and during the morning each day they towed a tea-trolley into the hop garden and sold cakes and buns and mugs of tea for a halfpenny. Their surgery was opened during the morning for an hour for special cases, and after that the group went back into the hop garden to chat with the hoppers and help them at the bins. In the evening, surgery was opened again, and they manned a canteen. They had a mobile dispensary which motored round neighbouring farms and also served as an ambulance to rush patients to hospital at Pembury or to the Little Hoppers' Hospital at Five Oak Green.

The Crowhurst Farm mission was one of many all over the Weald at this period. They all had their own particular routines and special features, and there was even a certain amount of rivalry on their behalf among the hoppers on the different farms who boasted of the superiority of 'their' mission over all others.

Most of them held church services, often in the open air, as part of their activities, and organized Sunday School for the children, and this aspect of the missions was more or less prominent according to the mission and the people in charge of it but, in general, they tried to avoid the old 'pound of tea religion' of a generation earlier, and were better-liked for it.

All missionaries had to be equipped to deal with a wide variety of ailments and misfortunes. As well as the inevitable cuts, stings, bites, scalds and burns, there were occasional cases of paratyphoid, pneumonia, diphtheria, toothache and miscarriages—all calling for a good deal more than simple first-aid skills. Diarrhoea and stomach upsets of various kinds were common, and there was always the possibility that apparently common symptoms apparently brought on by unfamiliar diet, unhygienic cooking practices or over-indulgence in 'scrumped' fruit from local orchards, might prove to be a much more serious case of, perhaps, food poisoning, or caused by eating poisonous berries.

There were, too, well-known problems peculiar to the hop gardens: hop rash (caused by the irritative effects of the bines on skins); hop-eye (inflammation caused by the chemicals with which the hops were dusted); and hop-wrist (a particular occupational hazard caused by the special wrist action used in picking hops).

Many of the missions regularly included among their staffs medical students and qualified nurses who, as well as being able to put their learned skills to good practical use looked upon a season in the hop gardens as a valuable way to widen their experience.

The British Red Cross Society first came into the hop gardens in Kent in 1922, when the County of London branch joined the Kent branch and together they opened dispensaries in the Tudeley and Yalding areas. It was felt that, as most of the pickers in that area came from London, it was only reasonable that the London BRCS volunteers should look after them and help relieve the annual pressure on local doctors.

Work began in a tent at Tudeley. The tent was later replaced by a hut, which was later enlarged with financial help from local farmers. Subsidiary dispensaries were staffed at Hartlake (Hadlow), Stilstead and Finches Farms, Capel. Other centres were at Yalding, Hunton, East Farleigh and Wateringbury, all staffed by BRCS volunteers and operating dispensaries at Collier Street, Buston Manor and West Farleigh.

The Kent Branch BRCS annual report for the hopping season of 1938 tells of the Branch serving two areas, East Kent from Canterbury and West Kent from Cranbrook. In that year, the BRCS had twenty dispensaries, and the report explained that work in the hop gardens was undertaken by the Society at the invitation of organizations already working in the areas. In the Canterbury area, for instance, the BRCS worked with the Church of England Temperance Society, which provided part of the equipment necessary for the dispensaries. The BRCS officers and members were, of course, all volunteers who actually paid for the privilege of doing the work they did—twelve and sixpence messing charges—and they lived under canvas.

All through the war the BRCS maintained a presence in the hop gardens although the London members thinned out a great deal. They were all engaged in First-Aid Posts in London itself, and the Kent Branch took over the Yalding, Wateringbury and East Farleigh centres from them. The London branch kept a skeleton staff at Tudeley and Hunton. As well as all the usual hopping ailments and emergencies, injuries caused by air raids were added to the work of the nurses and not only were all the available dispensaries in use but in some rural areas the war-time First-Aid Posts were brought into service for the hop-picking duration as well.

In 1945 mobile dispensaries were introduced at Benenden and Goudhurst and they remained a feature of Red Cross work among the hoppers after that. The peak of Red Cross activity among the hoppers was reached in 1950 when volunteers staffed 31 dispensaries and dealt with 4,452 cases, of whom 367 were referred to doctors and 48 were admitted to hospitals.

The number of dispensaries actually went higher than the 1950 number—up to 45 in 1954—but the number of patients had dropped that year to 2,899. Even so, as the annual report for that year commented, one in three hoppers had occasion to visit a dispensary during the season.

The report was also able to note, however, that year: "It is interesting to note that owing to the increased health of the nation and better living conditions, such diseases as scabies and impetigo have almost disappeared."

The British Red Cross Society continued its work among the Kentish hop-pickers right up to the time the machines ended the old-style hopping. In 1959, the Society's Kent Branch annual report commented: "Hopping as we know it has almost ceased to exist. Staff this year augmented by 27 Voluntary Members, and we were able to visit farms in areas covering Selling, Ashford, Tunbridge Wells, Maidstone, Tonbridge and the Medway Towns. Three dispensaries were open daily, at which 704 patients attended with a total of 1,384 visits."

As well as treating all the illnesses and ailments of the hoppers, the Red Cross workers also provided soup and stew dinners, a 'Red Cross special' rice and treacle sweet, as well as tea and buns.

The first real attempt to organize the various voluntary bodies into a co-ordinated corps came with the creation of the

Hop Pickers' Medical Treatment Board, which held its first meeting at the Sessions House, Maidstone, on Friday, June 6, 1930. The Board was an offspring of the Kent Rural Community Council (later to become the Kent Council of Social Service and today Kent Voluntary Service Council, based at Folkestone).

The chairman of the Health Committee of the Kent Rural Community Council was an *ex officio* member of the Board, the first chairman of which was Colonel A. S. Barham, Chairman of the Kent County Health Committee. The Board invited Lieutenant-Colonel A. Chopping to become its first Honorary Medical Adviser, whose role was to visit all the many field dispensaries operated by various voluntary organizations, and to report on the adequacy or shortcomings of them all. Dr Chopping, however, was unable to accept at once and the Board appointed Dr C. Ponder, Assistant County Medical Officer of Health to act as his deputy for that first year.

The Board had twelve founder member organizations. These were: The British Red Cross Society; Church of England Mission to Hop-pickers; Church of England Temperance Society Mission to Hop-pickers; Crowhurst Farm Mission; Hop-Pickers' Hospital and Medical Centre, Marden; Hop-Pickers' Mission for the Weald of Kent; Horsmonden Medical Mission to Hop-pickers; Little Hoppers' Hospital at Five Oak Green; Mereworth Voluntary First Aid Centre; Salvation Army; Staplehurst Dispensary; and Stilstead Farm Dispensary, East Peckham.

All of these had begun work on an *ad hoc* basis, and some had been active among the hop-pickers for many years, and had considerable experience of the problems and needs.

In later years, the original twelve were to be joined by others, and there were several changes over the years, although the total number remained more or less constant throughout the life of the Board.

Some of the voluntary dispensaries had been charging their patients small sums of money for the treatment they gave, and there was talk of introducing an overall scale of charges, but this was decided against. It was left to the different organizations to make their own rulings about that.

The Hop Pickers' Medical Treatment Board was a semi-official (at least, officially recognized) body, and its influence was felt almost from the very beginning. Only two years after its formation, Colonel Chopping, who by this time had taken up his duties as Honorary Medical Adviser and was making his annual visits to the hop gardens and reporting to the Board on what he found, was able to report "great improvements". In 1930, he recalled, there had been acute difficulty in transporting patients to hospitals when it was necessary. By 1932, though, that problem was largely overcome. There were more trained nurses on duty at the medical huts and there were more calls upon the services of the District Medical Officers, too.

In 1932, Colonel Chopping noted there were two hospitals, and 47 dispensaries meeting the needs of hoppers in 237 camps. During the season, 6,787 new patients were treated. Between them, they made 13,461 attendances at dispensaries, and 86 patients were transferred to hospital for treatment. They were treated by five doctors, four medical students, 38 trained nurses, 75 part-trained nurses and 29 other volunteers.

Although the staffing was no great problem financially, the medical supplies had to be bought and the Board looked around for possible sources of extra income, above the £25 it received from Kent County Council. It hit upon the idea of asking Lord Cornwallis to broadcast an appeal for money by BBC, and this was done in August, 1933. It resulted in about 300 donations totalling about £320 but ranging from individual sums of sixpence to donations of £100. The money was shared out among member organizations.

In 1933, Colonel Chopping reported that, despite the dry and generally healthy season, there was a large increase in the number of patients treated, and he suggested it might have been because pickers were getting more used to using the dispensaries and were less shy about attending.

In 1936 another BBC broadcast appeal (on August 9), this time by Lord Harris, realized just under £100. The appeals were by now a regular feature of BBC broadcasting, and when the Board enquired about future appeals, members were told there was a long queue.

In fact, it was not until 1947 that the next appeal was broadcast for the Board and then 197 donations brought in almost £150.

In 1938, Colonel Chopping resigned as Hon. Medical Adviser, and the war years disrupted the Board's activities so that it was not until 1946 that his successor, Colonel J. H. Campbell, Senior Assistant County Medical Officer of Health, took over the job. From that time on, the work of the Board and its constituent members increased considerably.

In 1947 there were 18 dispensaries in the Mid-Kent and Weald area, and 10 in East Kent, all staffed by British Red Cross Society personnel. The St John Ambulance Brigade staffed 12 more in East Kent, and the Church of England (Paddock Wood) Mission operated a mobile dispensary that visited 15 camps in that area. The Cambridge Hopping Mission had 4 dispensaries in the Goudhurst and Cranbrook area; the Crowhurst Mission Association served 11 camps in the East Peckham area from two static and one mobile dispensaries; and the Trinity College, Oxford, Mission had a number of medical students working in one main and two smaller camps at East Peckham. The Horsmonden Hop-pickers' Medical Mission staffed one dispensary and looked after more than 20 camps, and the Little Hoppers' Hospital also had 3 dispensaries in the Five Oak Green and Paddock Wood area.

Near Pembury, St Colomba's Mission to Hop-pickers looked after three large hoppers' camps, and the Salvation Army had 7 dispensaries serving about 22 camps between Maidstone and Goudhurst. In addition to all these, there were also dispensaries at Laddingford and Collier Street (Yalding) which had been established and were maintained by local growers at their own expense.

Altogether, in 1947, there were 68 static first-aid centres and one mobile unit. Together, they served about 320 camps, from which 10,612 patients were treated. Those patients made 23,100 attendances altogether during the season. About 560 of them had to be referred to local doctors for National Health Service treatment, and 130 were transferred to general hospitals.

Much more detailed interest was being taken by now in the general health of the nation, and the hoppers' illnesses came under special scrutiny. It was found that, during the season, there were 633 cases of the peculiar hopping ailments of hop-eye, hop-ear, and hop-gout. Of all cases treated, 1,747 were accident victims of one sort or another.

The finances of the Hop Pickers' Medical Treatment Board took a turn for the better in 1949 when the Kent County Council found itself able to make a grant of £550 to the 11 voluntary organizations making up the membership. The grant, which was specifically only for the health services provided to the hoppers by the organizations, was described as "for those parts of their services which could be regarded as coming within the Council's responsibilities under the National Health Service Act, 1946". With the grant came an invitation to any of the organizations to ask for more money if they found themselves in financial difficulties over providing these particular services. From that time on, the Board received its grant every year and ceased to be a wholly voluntary organization in the same sense that it had been before then, almost to the end.

In 1951 the Board got its third Honorary Medical Adviser. Colonel Campbell retired and was succeeded by Dr J. H. Hazeldene, Senior Assistant County Medical Officer. He it was who saw the introduction of hop-picking machinery. In his first report, on the 1951 season, Dr Hazeldene commented:

> "The Hop Industry Productivity Team's recommendations for mechanisation of hop-picking would, as yet, appear to have made little impact on the traditional English method. Mechanisation was tried by less than a dozen experimental machines. There will be need for the voluntary dispensaries to provide first aid for the hoppers and their families for many years."

In fact he underestimated the speed of change. Within ten years, the 'traditional English method' was virtually gone from the Kentish hop gardens altogether.

The new National Health Service meant that hoppers could seek medical advice and treatment, free, from local doctors during their stay in Kent. But many still preferred to make the familiar and much more informal local voluntary dispensaries their first call for aid. It was, perhaps, just as well that they did. That way, the dispensaries were able to act as a filter, treating such cases as they could and passing on to the doctors only those that really did seem to need more professional treatment.

Besides, local doctors could well have been thrown into complete confusion if too many of the temporary visitors had gone directly to the surgeries. As it was, the Londoners caused quite a lot of difficulty by getting treatment from one local doctor and then changing to another for follow-up treatment without going through the proper processes for changing doctors.

In 1952, more than 37,000 hoppers in 230 camps provided 11,000 patients for the 10 voluntary organizations still working in the Kent hop gardens. Even as late as this, there were causes for complaints about conditions on some of the farms. Fresh water was not invariably available, and there were fears of possible outbreaks of gastro-enteritis but, in general, the health of the hoppers and the conditions in the hopper camps were both vastly improved compared with those of pre-war days.

Throughout the second half of the 1950s, the hop-picking machines increased in number and this, as much as the general improvement in health, lightened the work-load of the volunteers. In 1955, Trinity College, Oxford, abandoned its work at Stilstead Farm, East Peckham because of the effect of the introduction of machines, and became the first of the voluntary organizations to withdraw.

The effect of the machines on the dispensaries was, though, double-edged in one way. The machines brought new elements into the case-books, with accidents caused by the machinery, and also with cases of tenosynovitis, first reported in the hop gardens in 1961. This was an inflammation of the tendons of the forearm, brought on by the repeated movements necessary to extract unwanted fragments from the machine belt. It was, in fact, the 'mechanized' version of the old hop-gout, which settled into the wrists of hand-pickers.

But the peak was passed for the voluntary organizations. In 1953 there had been 81 dispensaries serving 223 camps; in 1954, 80 dispensaries served 239 camps. By 1961 the number of dispensaries was down to 23, and the number of camps down to 47. Far more than 26,000 attendances at the dispensaries in 1951 had dwindled, by 1961, to barely 3,000. Only six of the original voluntary bodies were still active in the hop gardens.

And so, in 1964, the Hop Pickers' Medical Treatment Board held its last meeting. Captain H. G. Giffard presided over the dissolution of its little "empire". In his report to the members, Captain Giffard said that the Manager of the Hops Marketing Board had regretted his Board was unable to give any financial help, and the Kent Council of Social Service had decided that, owing to the great reduction in the number of hop-pickers, there was no longer sufficient co-ordination work to be done to warrant KCSS participation. The KCC grant, which from 1949 until 1960 had been £500, had been gradually reduced thereafter until, now, it had ceased altogether.

The minutes of the meeting follow this report with the single sentence: "The meeting then agreed that the Board should be dissolved."

The formalities ended with the chairman proposing from the chair that there should be no further meeting of the Board, which should be dissolved as soon as the trustees had wound up its affairs. The resolution was passed unanimously.

SOURCES

The Hop Industry, Hubert H. Parker
Hops, A. H. Burgess
George Orwell in Kent, George Orwell
English Hops Glossary, Cronk
Treatise on the Cultivation of Hops, H. M. Manwaring
The Hop Industry, Anglo-American Council on Productivity
English Hops, George Clinch
Among the Hop-pickers in the Weald of Kent, Rev. R. Shindler
Hops and Hop-pickers, Rev. J. Y. Stratton
Whitbread Craftsmen, The Whitbread Library
Chambers' Book of Days (II, p. 399)
Life in Rural England, W. C. Finch
Nooks and Corners of English Life, John Timbs
The Rural Landscape of Kent, McRae and Burnham
Portrait of Canterbury, John Boyle
Among the Hop-pickers, Rev. Samuel Chinn
Kent, Richard Church
A History of Kent, Frank Jessup
Essays in Kentish History, Margaret Roake and John Whyman
Kentish Sources: *Aspects of Agriculture and Industry*, Kent County Archives
St Francis of the Hopfields, Miles Sargent
The Kentish Garland, De Vaynes
Cobbett's *Rural Rides*
History of the Kent County Council, 1889–1974, Elizabeth Melling
Old Days in the Kent Hop Gardens, (Ed.) Mary Lewis
Hop Tokens—Kent and Sussex, Rev. R. W. H. Acworth
Kent Voluntary Services Council (Minute books of Hop Pickers' Medical Treatment Board)
Annual Reports of Kent County Medical Officers
Hops Marketing Board, Paddock Wood
Ministry of Agriculture, Fisheries and Food
British Rail Archives

Kent Messenger Archives
Stede Hill, Robert H. Goodsall
Annual Reports, British Red Cross Society
Church Army Archives

INDEX